Encyclopedia of Alternative and Renewable Energy: Wind Power Systems Modeling

Volume 18

Encyclopedia of Alternative and Renewable Energy: Wind Power Systems Modeling

Volume 18

Edited by **Benjamin Wayne and David McCartney**

New York

Published by Callisto Reference,
106 Park Avenue, Suite 200,
New York, NY 10016, USA
www.callistoreference.com

Encyclopedia of Alternative and Renewable Energy:
Wind Power Systems Modeling
Volume 18
Edited by Benjamin Wayne and David McCartney

International Standard Book Number: 978-1-63239-192-6 (Hardback)

Printed in the United States of America.

Encyclopedia of Alternative and Renewable Energy: Wind Power Systems Modeling
Volume 18

Edited by **Benjamin Wayne and David McCartney**

New York

Published by Callisto Reference,
106 Park Avenue, Suite 200,
New York, NY 10016, USA
www.callistoreference.com

Encyclopedia of Alternative and Renewable Energy:
Wind Power Systems Modeling
Volume 18
Edited by Benjamin Wayne and David McCartney

International Standard Book Number: 978-1-63239-192-6 (Hardback)

Contents

Preface

Every book is initially just a concept; it takes months of research and hard work to give it the final shape in which the readers receive it. In its early stages, this book also went through rigorous reviewing. The notable contributions made by experts from across the globe were first molded into patterned chapters and then arranged in a sensibly sequential manner to bring out the best results.

This book encompasses the latest development and advancement of the wind energy conversion system. It consists of information contributed by renowned researchers from across the globe associated with the field of wind energy. The book highlights grid integration issues, dynamic and transient stability studies, and contemporary control theories applied in wind energy conversion system. Various other topics like modeling and control methodologies of distinct variable speed wind generators like switched reluctance generator, doubly-fed induction generator, permanent magnet synchronous generator, including the suitable power electronic converter topologies for grid integration, are also elucidated in this all-inclusive book. Real-time control analysis of wind farm with the help of Real Time Digital Simulator (RTDS) is also illuminated, along with Fault ride through integrated power flow solutions, wireless coded deadbeat power control, street light application, direct power control, etc.

It has been my immense pleasure to be a part of this project and to contribute my years of learning in such a meaningful form. I would like to take this opportunity to thank all the people who have been associated with the completion of this book at any step.

Editor

Dynamic Characteristics Analysis of Wind Farm Integrated with STATCOM Using RTDS

Adnan Sattar, Ahmed Al-Durra and S.M. Muyeen

Additional information is available at the end of the chapter

1. Introduction

This work concentrates on design and analysis of STATCOM connected at the wind farm terminal in real time environment using Real Time Digital Simulator (RTDS). This work is a part of power hardware-in-loop (PHIL) test required in a future project, and therefore, individual components are models in such a way that is close to real system. For the sake of detail analyses and future study, the system is simulated in two ways. First method is a dual time step approach, where wind turbines and generators of a wind farm, power grid, and control system are realized in the large time-step main network, however, 2-level voltage source converter based STATCOM is modeled in RTDS small time-step environment to adapt with higher switching frequency, where interface transformer is used to link the different time step sub-networks. In the second method, the entire system including the STATCOM is simulated in large time step. Detailed switching scheme for STATCOM and control strategy for both methods are discussed. An option for integrating anemometer for dynamic characteristics analysis is kept open, difficulties of STATCOM switching schemes for control prototype and PHIL testing in RTDS environment are discussed. The merits and demerits of both methods are also presented which is one of the salient features of this study. Results of RTDS are compared with Laboratory standard power system software PSCAD/EMTDC and the features of using RTDS in dynamic characteristics analyses of wind farm are also discussed.

2. Real Time Digital Simulator (RTDS) — A brief overview

2.1. Hardware

Real time digital simulator hardware is based on the parallel processing architecture and has been designed specially to solve the electromagnetic transient simulation algorithm. RTDS simulator consists of multiple RACKs, each of which consist of both communication and processor cards and are linked by a common backplane. To solve a large power system network, it is possible to split the entire power system into parts and these parts can be solved on the different subsystems or even using different racks on the RTDS simulator. Each rack has an Inter Rack Communication (IRC) card which allows the information to be shared between the different racks of RTDS. This study is carried out on RTDS consist of 3 RACKs.

Each rack has also a Workstation Interface (WIF) card which synchronizes the simulation calculations and communicates between different processor cards, as well as communication between different racks of the RTDS simulator. Also WIF card provides Ethernet communication to and from the graphical user interface during real time simulation.

The processors cards are responsible for the calculation of complete network behavior. RTDS uses two different processor card, 3PC (Processor card) and GPC (Gigabyte processor card). GPC contains 2 RISC processors running at 1GHz. Due to their computational power, they are often used in more than one component model calculation at the same time. It is noted that PB5 processor card, the next generation of GPC card is available in market from 2011, which has additional computation power and communication flexibility. Besides that, RTDS has a family of GT I/O cards. They are used with the GPC cards. GT I/O cards include analogue and digital input and output with 16-bit data converters. Other physical devices can be connected to the RTDS hardware by GT I/O cards.

2.2. Software

The graphical user interface between RTDS hardware and user is done by its own software, called RSCAD. It allows simulation circuit to be constructed, run, operated and results to be recorded. The RSCAD has 2 main modules, the Draft and the Run time. In Draft, an extensive library for both power system and control system components is available. The circuit can be constructed by copying the generic components from the library. After completion of the circuit, it will be complied in order to create the simulation codes required by the RTDS simulator. The simulation can be run using RSCAD Run Time module. Run time, operates on a PC or on workstation, back and forth communication with the WIF card through Ethernet. Simulation result can be plotted and operating condition of the system can be changed in run time by using switches, push buttons, etc., like the real world electric control rooms. A special module exists in RSCAD, so called T-LINE module, facilitate entry of transmission line data. Input information is related to the line geometry and conductor type. Multi-plot is used to analyze the graphical results and also to prepare it in report ready format. Several functions are available e.g., Fourier analysis and Total Harmonic Distortion computation. Figure 1 and 2 shows the RTDS hardware and RSCAD software modules [1-4].

3. Model system

The model system used for the simulation is shown in Figure 3. Aggregated model of the wind farm is considered in this study in which many wind generators in a wind farm are represented with a large wind generator. The induction generator is connected to the grid through the step up transformer and double circuit transmission line. Realistic data for the transmission line is used which is calculated from the transmission line length. As transmission line length is very important because of the STATCOM voltage support set point is considered at the common coupling point [5].

Figure 1. Real Time Digital Simulator (RTDS) Racks Installed in Electrical Department, The Petroleum Institute.

Figure 2.

(a)

(b)

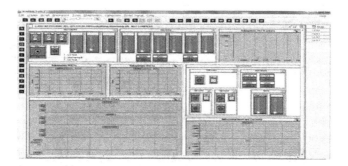

(c)

Figure 2. (a): RSCAD Software Modules. (b): Draft Module. (c): Run Time Module.

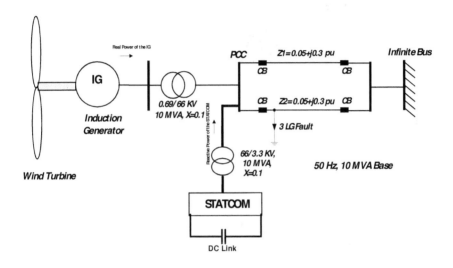

Figure 3. Fixed speed WTGSs including STATCOM connected at the PCC.

4. Real time simulation setup

Figure 4 shows the real time simulation setup required for this study. As discussed earlier, RTDS can be connected with other physical devices, an anemometer is considered to measure the wind speed data from the real site which will be sent to the RTDS via GT I/O cards. The wind speed signal will then be sent to RSCAD environment though workstation interfacing card and will be used in wind turbine model to produce torque for wind generator. Hence, this is the most accurate way of analyzing the behavior of wind turbine generator behavior at different operating conditions.

Figure 5 shows the RSCAD model system. The Induction machine is driven by the fixed wind turbine and is connected to the electric grid through the step up transformer and double circuit transmission line. The STATCOM is connected at the high voltage side of the transformer.

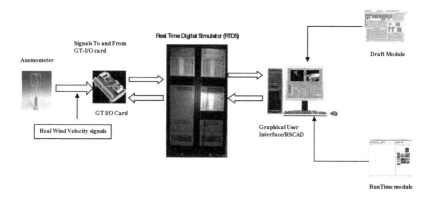

Figure 4. Real time simulation block diagram.

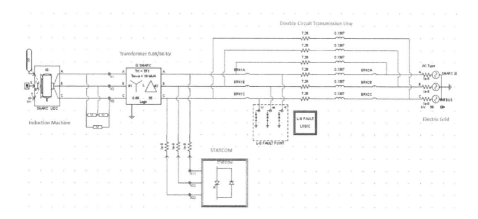

Figure 5. Fixed speed WTGSs including STATCOM connected at the PCC.

5. Modeling Of STATCOM

The modeling of STATCOM is completely done in the RTDS environment in two different methods, one in the dual time-step approach and second in the large time-step approach of the RSCAD. The details of both methods are presented in the following sub-sections.

5.1. VSC large time-step modeling of the STATCOM in RTDS

The power system components and control system components are modeled in the large time-step environment. Large time-step network solution is 50 μsec. STATCOM is also modeled in the large time-step environment and then coupled with the rest of the system through the transformer at PCC. Figure 6 shows the RTDS modules and processor assignments in large time-step approach. The power system components are solved on the 3PC card, while the control block is solve on the GPC card. GTO model is computed on the 3PC card. The switching is done on the GPC card.

5.2. Switching scheme in large time-step

For the large time-step, switching is done by using the pulse width modulation (PWM) technique. For GTO model, switching signal can be generated in two ways either by using one 6P Grp which will generate 6–bit firing pulse integer word (FP), one active bit integer word (FLAST) or by 3 LEG mode. In 3 LG mode, there is one 2-bit firing pulse word, one active bit word and one fraction word for each separate leg in the valve group. In this work, switching is done by using 3 LEG mode. The switching diagram is shown in the Figure 7.

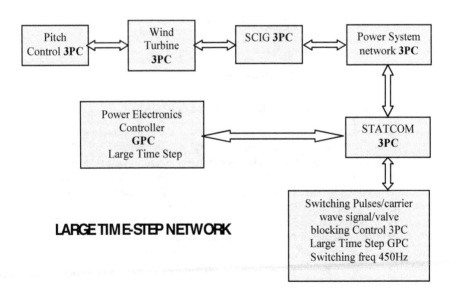

Figure 6. RTDS modules and processor in large time-step approach.

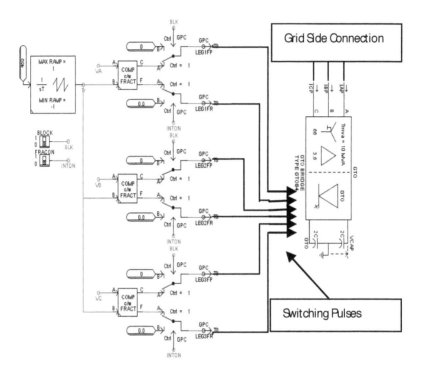

Figure 7. Switching scheme for STATCOM in large time-step approach.

5.3. VSC dual time-step modeling of the STATCOM in RTDS

In this thesis, the STATCOM model is also developed in the small time-step environment of the RTDS. Power system components and control system components are modeled in the large time-step environment. Thus model system is run in two different time step, small time-step normally run at 1μsec – 4μsec and large time-step typically running at 50μsec. Thus two different time-step simulation is interfaced each other through the interface transformer. RTDS modules and the processor assignments have been shown in Figure 8. Main power system components are solved on the 3PC card. The STATCOM has been modeled in the Voltage Source Converter (VSC) small time-step network, which are solved on the GPC card. The control part is solved on the 3PC card. Carrier wave signal is generated in the large time-step and are imported in the small time-step after being adapted with small-time step. The carrier wave frequency is chosen 2 kHz.

5.4. Switching scheme in dual time-step

As mentioned earlier that dual time-step operates in both small and large time steps. Small time-step VSC sub-network model including switching scheme is shown in Figure 9. Using the principal of pulse width modulation scheme, the carrier and modulation signals are generated in the RTDS large time-step size environment and then processed to generate high resolution firing pulses using the RTDS firing pulse generator component in the small time-step environment. In order to ensure accurate firing this component requires the transfer of reference phase and frequency from the large time-step environment. This allows the component to extrapolate the phase between large time-steps. The valves of the GTO bridge gets firing pulse input from the comparator by selecting the option CC_WORD of GTO bridge. The valves of the GTO bridge are controlled by the respected bits in a firing pulse word. These consecutive bits are aligned in such a way that the least significant bit (LSB) in the firing pulse coincide with the LSB in the final applied firing pulse word. Hence, the first LSB controls the valve 1, the second LSB controls the valve 2, and so on [6-7].

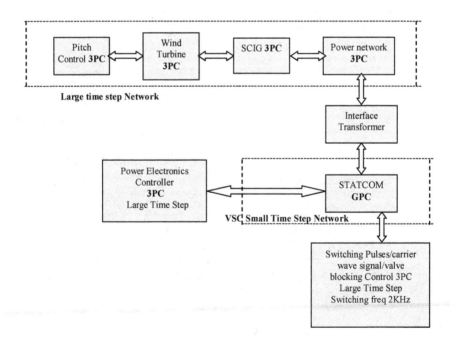

Figure 8. RTDS modules and processor in dual time-step approach.

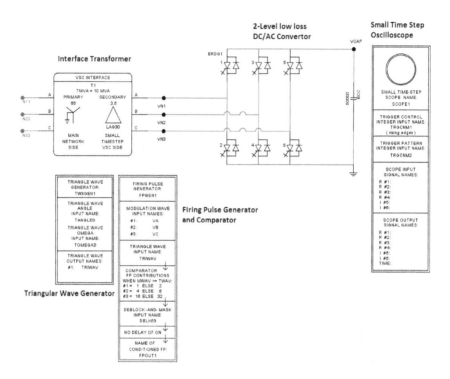

Figure 9. Switching scheme for STATCOM in small time-step VSC sub-network (part of dual time step approach).

6. STATCOM control strategy

The cascaded vector control scheme is considered for the control of the STATCOM, in this study. The control block diagram of VSC based two level STATCOM is shown in Figure 10. The aim of the control is to maintain desired voltage magnitude at the wind farm terminals during normal operating condition and recover the voltage in shortest possible time after occurrence of grid fault. The current signals are measured at the high voltage side of the transformer and are these three phase quantities are transformed in to two quantities i.e. I_d and I_q by the abc to dq Transformation. DC link voltage and the terminal voltage is controlled. The outer loop will generate the reference signals for the dq quantities and inner loop will keep the system to its desired output. The two voltage reference signals V_d and V_q are generated and are transformed to the three phase voltage reference signals for the switching V_a, V_b and V_c by the dq to abc transformation. The reference transformation angle used for the abc-dq conversion is generated by the three phase grid voltage signals by using the Phase Locked Loop (PLL).

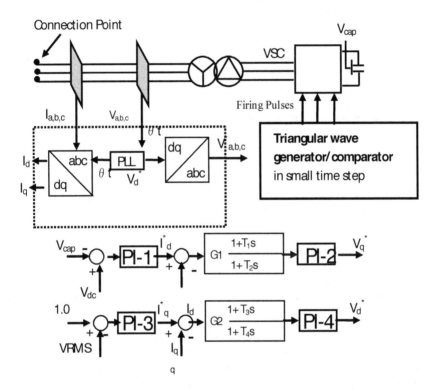

Figure 10. Control block diagram of VSC based STATCOM.

With suitable adjustment of the magnitude and phase of the VSC output voltage, an efficient control of power exchange between the STATCOM and the ac power system can be obtained. The vector control scheme generates the three-phase reference signals which are used to generate the switching signals for the GTO switched STATCOM. The STATCOM rating has been considered as the same of wind farm rating. The rated DC link voltage is 6.6 kV. The STATCOM is connected to the 66 kV line by a single step down transformer (66 kV/3.6 kV) with 0.1 p.u leakage reactance. The DC-link capacitor value is 50000μF. The values of the PI controller used are set by the trial and error method to get the best results [8-9].

7. Simulation results

In this paper dynamic characteristic is analyzed when STATCOM is considered to be connected at wind farm terminal. Keeping in mind the future control prototype and PHIL testing, STATCOM is modeled in both dual and large time-step environment. Real wind speed data is measured, stored in data file, and used in RTDS environment using scheduler which will finally be replaced with advanced anemometer equipped with remote data logger. Realistic data is used in transmission line calculated from transmission line length which can be changed suitably with any wind farm site data in the next step. Line length is important because STATCOM voltage support set point is considered at the common coupling point. Results are also compared with PSCAD/EMTDC where time step is considered as 20 sec and switching frequency is considered as 2000Hz. Detailed switching model is considered to model STATACOM in PSCAD/EMTDC environment to perform the time comparison.

7.1. Dynamic characteristics analysis

The analysis is carried out using 50 sec of wind speed data. Interpolation technique is not considered while using real wind speed data in the simulations using PSCAD/EMTDC and RTDS/RSCAD. Longer period can be considered based on the available memory resources. As the wind speed is changing randomly, In Figure 11, the important responses using offline simulator PSCAD/EMTDC are shown, when STATCOM is considered to be connected at wind farm terminal. The wind farm terminal voltage cannot be maintained at constant value using only the capacitor bank of rated capacity. When STATCOM is used, terminal voltage of wind farm can be maintained at the desired level set by Transmission System Operators (TSOs), as shown in Figure 11.

Figures 12 and 13 shows the responses obtained using RTDS in dual and large time-step approaches, respectively. To match the switching frequency used in PSCAD/EMTDC, in dual time-step approach 2000Hz carrier frequency is considered. In Dual time-step approach, VSC sub-network is simulated using 1.5sec and the other components are simulated in large time-step of 45sec as it is required to run the simulation higher than the suggested minimum time-step by RTDS resolver. On the other hand, in large time-step approach, the time step chosen is also 45sec for the sake of time comparison of both approaches. It should be noted that low switching frequency should be used in large time-step approach for generating the switching pulses for switching devices, which is considered as 450Hz in this study. In both figures, the step change around 10 sec represents the change of machine state from constant speed to normal operation. Figures 11 to 13 shows good agreement for IG real power, STATCOM reactive power, DC-link capacitor voltage, wind farm terminal voltage, and IG rotor speed responses, except the initial responses of first few seconds. However, dual time-step approach using RTDS gives smooth responses compared to large time step approach in RTDS and PSCAD/EMTDC, due the exact switching ability in the range of less than 2sec.

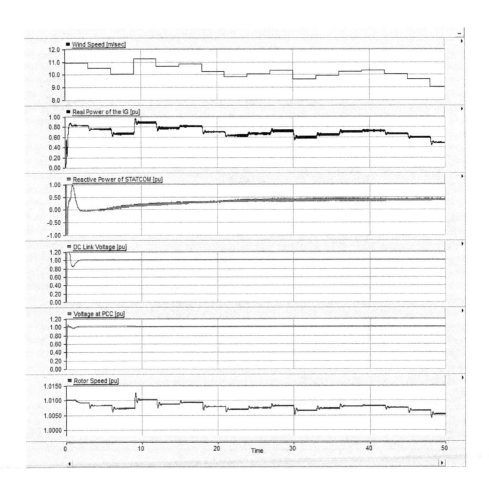

Figure 11. Dynamic characteristics responses obtained using PSCAD/EMTDC.

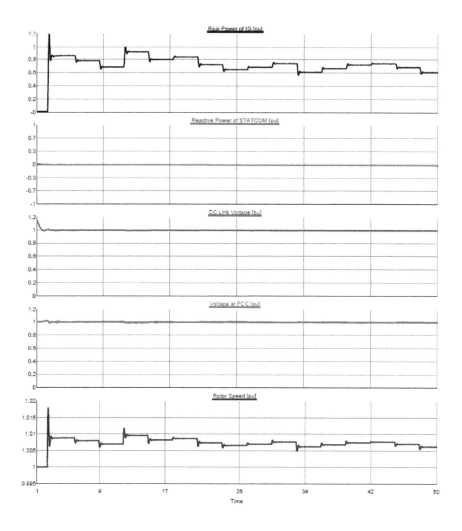

Figure 12. Dynamic responses obtained using RTDS (dual time-step approach).

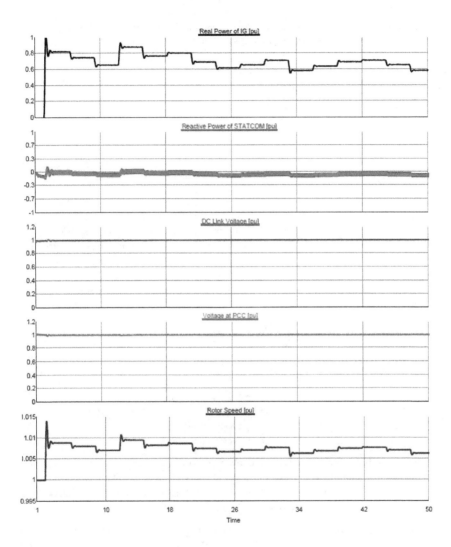

Figure 13. Dynamic responses obtained using RTDS (large time-step approach).

A time comparison is carried out while analyzing dynamic characteristics using 50 sec of real wind speed data using both PSCAD/EMTDC and RTDS/RSCAD. Both dual and large time-step approaches require almost the same time of about 51sec to download and plot the result. However, in PSCAD/EMTDC a total time of 720 sec is required to finish the simulation of 50sec, though the program is simulated using 20sec, which is lower than RTDS large time step. Table 1 shows the comparison between the times taken by the PSCAD/EMTDC and RSCAD/RTDS.

Therefore, it is quite difficult to perform dynamic analysis for longer time in the range of hour or day to determine the optimum capacity of STATCOM suitable for a real wind farm.

Simulation Technique	PSCAD/EMTDC	Dual Time-Step Approach	Large Time-Step Approach
Time in seconds	720	51	51

Table 1. Simulation time comparison.

8. Conclusion

In this study, a detail dynamic analysis of grid connected wind farm integrated with STAT-COM has been carried out using offline laboratory standard power system simulation tool PSCAD/EMTDC and Real Time Digital Simulator (RTDS). Detailed modeling, control, and switching scheme of STATCOM have been presented, suitable for wind energy conversion system. Dual (small and large) time-step approach and approach using large time step to simulate STATCOM in RTDS environment have been demonstrated. A comparative study has also been performed which are summarized as follows.

Offline simulation technique using PSCAD/EMTDC, MATLAB/Simulink, PSS are precise enough. However, the simulation takes much long time which is practically not feasible for the dynamic analysis in hour or day range, especially when detailed switching model is considered. RTDS is an effective tool for such type of analysis due to fast computation capability.

Dual time step approach is the most accurate method to simulate power converter in RTDS environment. Besides that, dual time step approach is also good to conduct the loss analysis of power converters operated at higher switching frequency.

The system including power converters can even be simulated using large time step, which requires almost the same time of dual time step. The large time step VSC bridge available in RTDS/RSCAD library has Digital Time-Stamp (DITS) feature to handle switching pulses from real world or external DSP/MATLAB based system.

RTDS resources can be used in optimum way simulating power converter using large time-step approach in 3PC card, when GPC card is fully utilized for dual time-step power converter simulation or other purposes.

Author details

Adnan Sattar, Ahmed Al-Durra and S.M. Muyeen

Electrical Engineering Department, The Petroleum Institute, Abu Dhabi, UAE

References

[1] Forsyth, P, & Kuffel, R. Utility Applications of a RTDS Simulator, " 2007 IPEC International Power Engineering Conference, Dec (2008). , 112-117.

[2] Kuffel, J, Giesbrecht, T, Maguire, R. P, Wierckx, P. A, & Forsyth, P. G. Mclaren, " RTDS- A Fully Digital Power Simulator Operating in Real Time, " 1995 WESCA-NEX Conference Proceedings on Communications, Power, and Computing, May (1995). , 300-305.

[3] Kuffel, R, Giesbrecht, J, Maguire, T, Wierckx, R. P, Forsyth, P. A, & Mclaren, P. G. A Fully Digital Real-Time Simulator for Protective Relay Testing, " 1997 Developments in Power System Protection, Sixth International Conference, Mar (1997). , 147-150.

[4] Real Time Digital Simulator Power System and Control User ManualRTDS Technologies, (2009).

[5] Muyeen, S. M. et al., "Stabilization of Grid Connected Wind Generator by STATCOM," *International Conference on Power Electronics and Drive Systems (IEEE PEDS 2005)*, Conference CDROM, Malaysia, (2005). , 1584-1589.

[6] Qi, L, Langston, J, Steurer, M, & Sundaram, A. Implementation and Validation of a Five-Level STATCOM Model in the RTDS small time-step Environment, " *2009 PES Power & Energy Society General Meeting,* Jul (2009). , 1-6.

[7] Wei QiaoGanesh Kumar Venayagamoorthy, and Ronald G. Harley, "Real-time implementation of a statcom on a wind farm equipped with doubly fed induction generators", *IEEE Transactions on Industry Applications*, (2009). , 45(1), 98-107.

[8] Saad-saoud, Z, Lisboa, M. L, Ekanayake, J. B, Jenkins, N, & Strbac, G. Application of STATCOMs to wind farms", *Proc. Inst. Elect. Eng., Gen., Transm., Distrib.*, (1998). , 145, 511.

[9] Muyeen, S. M, Mannan, M. A, Ali, M. H, Takahashi, R, Murata, T, & Tamura, J. Stabilization of wind turbine generator system by STATCOM", *IEEJ Trans. Power Energy,* (2006). , 126-B, 1073.

Chapter 2

Direct Power Control for Switched Reluctance Generator in Wind Energy

Tárcio A. dos S. Barros, Alfeu J. Sguarezi Filho and
Ernesto Ruppert Filho

Additional information is available at the end of the chapter

1. Introduction

Wind energy is one of the renewable energy must used actually. It is the kinetic energy contained in moving air masses. Its use occurs through the conversion of kinetic energy translation in kinetic energy of rotation with the use of wind turbines to produce electricity. To ensure the best exploration of wind energy is necessary the use of generators that take of this energy form more efficient in variable speed systems. The electrical induction machines and synchronous generators are widely used in wind turbines. The switched reluctance generator has been studied and pointed as a good solution for applications for wind generation systems of up to 500kW [1].

The switched reluctance generator (SRG) has as main characteristics: mechanical robustness, high starting torque,high performance and low cost. The SRG can operate at variable speeds and its operating range is broader than synchronous and induction generators. Some works that study the behavior of the SRG in case of variable speed are presented in [2–4]. In these studies, the control systems used for power control of SRG are controls that utilize pulse width modulation control and a current loop, but the use of these controllers has shown that their performances are not satisfactory due to switching losses and consequently decrease of efficiency in operations with varying speeds of SRG.

An alternative for the power control of electrical machine is the direct power control (DPC). This technique allows to control directly the power without the use current loops or torque and flux loops and it is applied satisfactory in DFIG and inverters [5] .

In this work is performed a control method of the switched reluctance generator through simulation techniques using mathematical models of the studied system. A wind power generation system with the SRG connected to the grid was performed based on control of two separate converters. The control of the converter connected to the SRG regulates the extraction of electrical power to be generated and the control of the converter connected to the grid is responsible for regulate the transmission of the generated energy to the grid. A direct power control was developed to control the power generated by SRG. Unlike most of SRG control schemes found in the literature in which the power of the SRG is indirectly controlled by a current loop, the direct control of power acts directly on the power generated by SRG. The energy generated by SRG is sent to the grid by a voltage source converter, which also controls the active power delivered to the grid.

2. Wind Energy Systems

From 1998 to 2008 the growth of wind power installed in the world was approximately 30% and in the last three years the value of installed wind power kept in the mean 45GW, with a total current value of 237.7 GW of installed power in the world. China has the largest installed wind power value (about 62.3 GW) followed U.S. (46.9 GW) and Germany (29.06 GW) [1].

The electrical machine widely used as a generator in wind power generation systems are the induction and synchronous [6, 7]. These generators can operate with variable speed depending on the use of electronic converters for the processing energy of generators. A machine that can be used in wind power generation systems is the switched reluctance machine [8, 9].

A schematic diagram of wind power generation system connected to the grid using the SRG is shown in Figure 1. This system is based on the control generation of two separate converters. The converter connected to the SRG regulates the maximum extraction of electric power according to the profile of wind system.

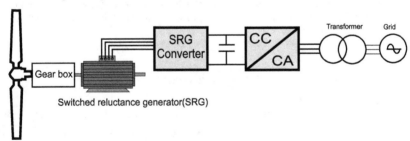

Figure 1. Structure of cascade converters for wind generation using the SRG.

In the literature were found articles that discuss the connection of SRG and the grid in wind power generation systems with variable speed. In [4] the authors used two strategies to control output power of a SRG 8/6. These experiments demonstrated a high efficiency of the system to a wide range of variation of speed. However, the PWM control used is contested by [10] for its hardware complexity for variable speed situations in a wide range of speeds. The converter used in [4] to drive the SRG uses a converter *buck* to magnetize the phases of the machine. This increases the cost and complexity of the converter.

In [2] was developed a system to control the power generated by the SRG using a hysteresis control. It was observed a satisfactory result just to low speed operation. In [11] has been proposed a control system where the power sent to the grid is controlled directly by the inverter connected to the grid. It was observed that this form of control has slow response and poor performance in situations of high speed variations.

A system that consists in controlling the power generated by a SRG 6/4 connected to a DC network has been proposed in [12]. The converter used requires a converter *buck-boost* voltage to regulate the magnetization of the SRG.

An alternative analyzed in studies in literature relates to the development of controllers to connect the SRG directly to the load to be fed through the converter SRG. In [13, 14] controls were performed using fuzzy logic to keep the generated voltage constant. Other controls using optimization of switching angle of of SRG were performed in [15, 16], but require high processing power and storage tables.

2.1. Mechanical power extracted from the wind

To use the energy contained in wind is necessary to have a continuous and fairly strong wind flux. The modern wind turbines are designed to achieve the maximum power in wind speeds in the order of 10 to 15m/s. The energy available for a wind turbine is the kinetic energy associated with a column of air that moves at a constant uniform speed. The mathematical model allows to calculate the aerodynamic torque mechanical value or mechanical power applied to the shaft of the electric generator from the information of the wind speed and position value of the step angle of the turbines. The model also depends on the type of the turbine to be represented as having the characteristics of vertical or horizontal axis, number of blades, blade angle control, and regardless of the type of electrical generator chosen or the type of control of converters. Accordingly, this allows it to be studied regardless of the types of electrical generators. The mechanical power in steady state can be extracted from the wind is shown in Equation 1 [17–19].

$$P_m = \frac{1}{2}\rho A v^3 C_p\left(\psi, \beta_t\right) \tag{1}$$

Since P_m mechanical power of the turbine, ρ density of air, A area swept by the turbine blades, v wind speed and C_P coefficient of performance, ψ a linear relation $\dfrac{\omega_r\,R_t}{V}$, R_t the radius of the turbine and β_t pitch angle of the vanes of the turbine. The power coefficient of C_P indicates the efficiency with which the wind turbine transforms the kinetic energy contained in wind into mechanical energy rotating. The power ratio depends on the linear relationship between the wind speed and the speed of the propeller tip psi and the propeller pitch angle β_t.

In expression (1) it appears that the mechanical power (P_m) generated by wind power directly depends on the power coefficient C_P. On the other hand, considering the pitch angle of the vanes fixed to zero, the power coefficient depends exclusively on the relationship between the velocity and the linear speed of the tip of the helix, therefore the mechanical power produced by a wind turbine is varied according to their operating speed. For each value of the wind speed is a region in which the rotor speed maximizes the mechanical power generated. Therefore, for wind speeds below rated speed operation with variable speed rotor increases efficiency in power generation [18, 19]. The profile of optimizing the efficiency of the power generated for variable speeds can be expressed by:

$$P_{opt} = k_{opt} w_r^3 \tag{2}$$

Where P_{opt} is the optimum power and k_{opt} depends of aerodynamics of the helix, gear box and parameters of the wind turbine.

3. Structure and operation principle of the switched reluctance machine

The switched reluctance machine (SRM) is a primitive conception, and its basic concept of operation has been established for around 1838. However, only with the development of the power electronics has become possible to use this machine operating system for applications requiring variable speed [20].

3.1. Basic structure

SRM is a double salient (in rotor and in stator) having field coils in stator as the DC motor and does not have coils or magnets in the rotor. The rotor is composed of ferromagnetic material with salient regularly. Figure 2 there is a MRV 8/6 (number of stator poles/rotor poles number). Other possibilities existing construction are 6/4, 10/4, 12/8 and 12/10, among other configurations.

The operation principle of the MRV as motor is based on the principle of minimum reluctance, ie, when the winding on a stator pole pair is energized, the poles of the rotor are attracted to a position that represents the minimum reluctance (axes aligned), generating a torque on the rotor. while two rotor poles are aligned with the stator poles other poles are out of alignment rotor. These other stator poles are driven bringing the rotor poles into alignment. By the sequential switching of the stator windings, there is production of electromagnetic torque and the rotor rotates [10].

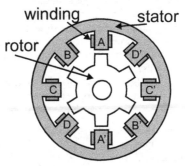

Figure 2. Front view of a switched reluctance machine 8/6.

4. The switched reluctance generator (SRG)

The switched reluctance generator (SRG) is an electromechanical energy converter capable of transforming mechanical energy into electrical energy. To operate as a generator, the machine must be excited during the degrowth of inductance and a mechanical torque must be applied on the machine shaft. The magnetization of the phase conjugate added to the input of the mechanical axis of the machine causes an emf appears which increases the rate of growth of the current curve [9, 21].

A typical drive system for switched reluctance generator is shown in Figure 3. This drive the SRG structure consists of a converter and a control system in closed loop since the SRG is unstable for operation in open loop [9]. The converter of Figure 3 drives the SRG and it is represented only one phase of SRG. Drive for a number of phase SRG higher is given below. The SRG can power the load directly as shown in Figure 3 or send energy to the grid using another power electronic converter.

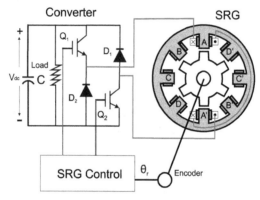

Figure 3. Drive system of SRG.

4.1. SRG converters

The SRG operates in two stages: excitation and generation. The excitation stage is performed when one phase of the SRG is submitted to the excitation voltage, which causes current flow in the winding of this growing phase (Figure (4)). In generating the current through the SRG phase to the load. In each period of the excitation voltage bus transfers energy to the magnetic field of the corresponding phase. In the generation period this energy flows to the load together with the share resulting from the conversion of mechanical energy into electrical [4, 20, 22]. Therefore, the SRG drive converter should be able to apply voltage in individual phases of the machine and create a path for the generated energy can flow to the electrical load.

There are several converters to drive the SRG, but the asymmetrical half-bridge converter AHB (Assimetric Half Bridge), Figure 5, is the most used because it allows robustness and the stages of regeneration energy and free-wheel when needed.

The converter of Figure 5 makes the SRG self excited, an initial excitation is required for the operation of the SRG. Usually the initial excitation is provided by an external source (a

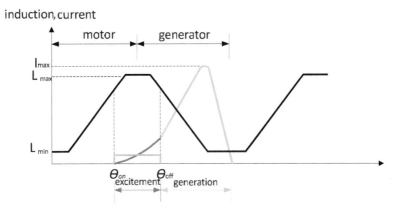

Figure 4. Profile of the Inductance..

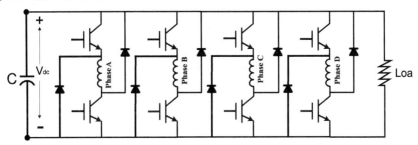

Figure 5. Converter AHB four phase.

battery for example) until the capacitor is charged. This pass the same capacitor exciting the phases when the external source is removed. The capacitor also has the function of stabilizing the voltage delivered to the load.

The Figures 6(a) e 6(b) illustrates the stages of operation of the AHB converter. In the excitation stage (Figure 6(a)) the two switches (of phase to be excited) turn on. Then a current flows from the capacitor to the stage, magnetizing it. After the excitation interval both switches are opened and the diodes are now to conduct the generated energy to the load and recharge the capacitor (Figure 6(b)). This process is repeated cyclically for each phase of the converter.

One of the main advantages of this converter is its flexibility in the control individually of current in each phase. Furthermore the converter not allowing short circuit in DC bus converter due to the fact that the switches connected in series with machine winding [10, 23]. However, this configuration is not the cheapest, because it requires two semiconductor switches per phase of the SRG. There are topologies that use less than two switches per phase of the GRV, as shown in [10, 20, 24], but these topologies have limitations regarding the AHB converter as the literature describes.

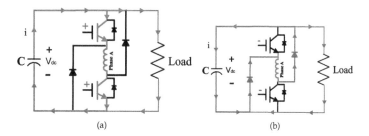

Figure 6. a) Excitation stage b) Generation stage.

4.2. SRG drive control

The SRG control is accomplished through the control of the switches of its converter. The SRG control requirements depending of each application. The quantities for controlling the generation of SRG is the period of excitement, the operating speed and excitation voltage [10]. When the load is connected directly to the converter (Figure 5) is necessary to provide a controlled voltage to the load. In case of load change control must act on the above quantities to maintain the generated voltage constant. This type of control is known as SRG voltage control bus. There are several controls proposed for this type of configuration as shown in [21]. This type of control is important for embedded applications such as vehicle and aerospace, where there is a need to maintain the dc bus voltage (responsible for food loads) at a constant value [10]. Figure 7 illustrates the typical configuration of this control.

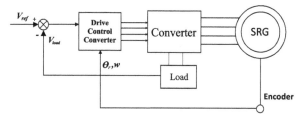

Figure 7. Bus voltage control structure.

When it is desired that the SRG operate at its optimum generation point is desirable controller the power generated directly [25, 26]. The typical control is shown in Figure 8. This control is typically used for power generation connected to the grid as seen in [25, 26]. The main applications of this type of control is in wind generation and generators driven by steam turbines . In this case the SRG is not connected directly with the load, but with another power converter that is responsible for sending the generated energy to the grid.

5. The power control system of the SRG

In this section its presents a technique of power control for the GRV connected to the grid. Unlike the controls proposed in [2, 4, 11] the converter connected to the generator is responsible for controlling the power to be generated and the other converter connected to the grid controls the voltage V_{dc} and its sends the energy generated to the grid.

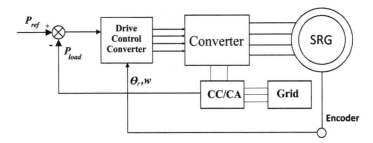

Figure 8. Generated power control structure.

5.1. Power converters

The converter used to drive the GRV was the AHB converter. This converter is connected via the DC link with the voltage source converter that it is connected to the grid.

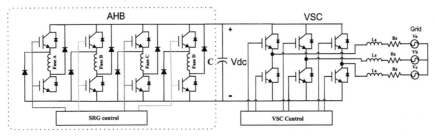

Figure 9. Power converters.

5.2. Direct Power Control

The direct power control system for SRG must regulate the power generated at the point of maximum aerodynamic efficiency, in other words, $P_{ref} = k_{opt}w_r^3$, where P_{ref} is the reference of active power. The proposed direct power control system consists in control the power generated by SRG directly. The diagram of the direct power control is shown in Figure 10. The control consists in keeping angle of activation of the switches of the HB converter at a fixed value θ_{on}. The PI controller processes the error (e_P) between P_{ref} and the generated power P and controls angle of shutdown of the switches θ_{off}, as shown in Equation (4). Thus, the principle of the control is when the step of the excitation increases, the generated power increases, as well.

The expression for the error power is given by:

$$e_P = P_{ref} - P \tag{3}$$

The angle θ_{off} is given by:

$$\theta_{off} = K_p e_P + K_i \int e_P dt \tag{4}$$

where: K_p is the proportional gain and k_i is the integral gain of PI controller.

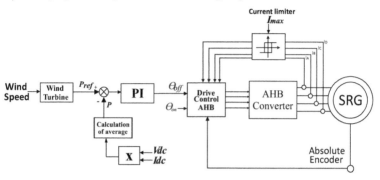

Figure 10. Diagram of direct power control for SRG.

5.2.1. Tuning of PI controller gains

The PI controller gains were adjusted using the second tuning method of Ziegler-Nicholds described in [27]. Figure 11 illustrates the procedure performed. Initially, the integral gain is zero and sets the proportional gain value (K_p) for the closed loop system, until the point where the response of the system starts to oscillate periodically. This setpoint is known as the critical point, in which the period of oscillation is defined as the critical period (P_{cr}) proportional gain and proportional gain is defined as critical (K_{cr}). From the P_{cr} e K_{cr} PI controller gains are determined using the relationship shown in Equation 5. This technique allows a good initial value of line when you do not know the process of plant be controlled.

$$K_p = 0.45 K_{cr}$$

$$K_i = \frac{1}{2} P_{cr} K_{cr} \tag{5}$$

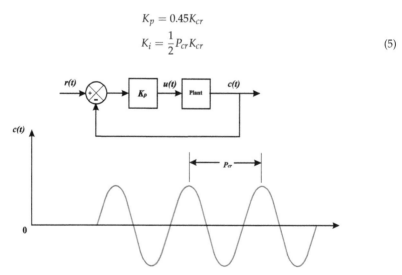

Figure 11. Second tuning method of Ziegler-Nicholds.

5.3. Power Grid Connection Converter

The Voltage Source Converter (VSC), shown in Figure 9, controls the voltage V_{dc} and it allows to send the generated power by SRG to the grid. The control strategy applied to the converter voltage source consists of two control loops. There is an internal control loop that controls the current sent to the grid, and, externally, there is a control loop of bus voltage control (V_{dc}). The current loop control (i_{sd}, i_{sq}) is responsible for controlling the power factor of the power sent to the grid [28]. The control voltage of the DC link is responsible for balancing the flow of power between the SRG and the grid [29].

The DC link voltage control of the voltage source inverter is realized in the synchronous coordinate system (dq) with employment grid voltage angle ($\theta = wt$) used in the transformation $abc\ dq$, which is obtained using a Phase-Locked Loop (PLL). The control voltage of the DC link (V_{dc}) is performed by a PI controller, which comes from the reference value i_{sd}^* (6), while the value of i_{sq}^* is derived from the power factor desired FP and P_{ref} (7).

$$i_{sd}^* = K_{pi}(V_{dc}^* - V_{dc}) + K_{ii} \int (V_{dc}^* - V_{dc})dt \qquad (6)$$

$$i_{sq}^* = \frac{-3}{2}\hat{P}_{ref}\frac{\sqrt{1-FP^2}}{FP^2} \qquad (7)$$

The reference values of current are compared with the values obtained from electrical grid (i_{sd} e i_{sq}) and are processed by two PI controllers that generate the value of the space vector voltage of grid $\vec{v}_{dq}$ (8) and (9) in the synchronous coordinate system (Equation 10). This space vector is transformed for the coordinate system abc generating the signals voltage $v_{mod_{abc}}$ (Equation 11) which are then generated using the PWM sinusoidal. The control system for VSC is shown in Figure 12.

$$v_{ds} = K_{ps}(i_{sd}^* - i_{sd}) + K_{is} \int (i_{sd}^* - i_{sd})dt \qquad (8)$$

$$v_{qs} = K_{ps}(i_{sq}^* - i_{sq}) + K_{is} \int (i_{sq}^* - i_{sq})dt \qquad (9)$$

$$\begin{bmatrix} i_d \\ i_q \end{bmatrix} = \frac{2}{3}\begin{bmatrix} cos\theta & sen\theta \\ -sen\theta & cos\theta \end{bmatrix}\begin{bmatrix} 1 & -\frac{1}{2} & -\frac{1}{2} \\ 0 & \frac{\sqrt{3}}{2} & -\frac{\sqrt{3}}{2} \end{bmatrix}\begin{bmatrix} i_a \\ i_b \\ i_c \end{bmatrix} \qquad (10)$$

$$\begin{bmatrix} v_a^{mod} \\ v_b^{mod} \\ v_c^{mod} \end{bmatrix} = \begin{bmatrix} 1 & 0 \\ -\frac{1}{2} & \frac{\sqrt{3}}{2} \\ -\frac{1}{2} & \frac{-\sqrt{3}}{2} \end{bmatrix}\begin{bmatrix} cos\theta & -sen\theta \\ sen\theta & cos\theta \end{bmatrix}\begin{bmatrix} v_d^{mod} \\ v_q^{mod} \end{bmatrix} \qquad (11)$$

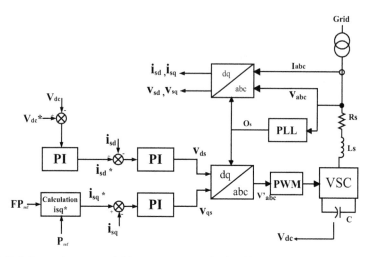

Figure 12. Block diagram of vector control of the converter connected to the grid.

5.4. PWM sinusoidal (SPWM)

There are several modulation techniques performed on the VSC to obtain a modulated voltage V_{abc}^{mod} in the terminals of the converter [30]. The most used technique is known as PWM sinusoidal (SPWM) using a triangular carrier for generating waveform desired [31].

SPWM modulation is obtained by comparing a reference voltage (sinusoidal) with a signal symmetrical triangular. The frequency of the triangular wave is to be at least 20 times greater than the maximum frequency of the reference signal, so that it is possible to obtain an acceptable reproduction of the waveform after filtering [32].

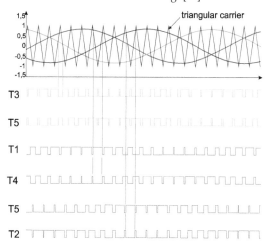

Figure 13. Modulation PWM sinusoidal

The Figures 13 shows the three-phase SPWM modulation signals used in this work. The carrier is compared with the reference sine wave for each phase, so that: when the carrier is greater than the reference phase, the switch top to respective phase is activated, otherwise the switch is operated below. Thus the output voltage of the modulated VSC is formed by a succession of rectangular wave. With a low pass filter can eliminate harmonic components generated by the modulation.

5.5. Synchronism with the grid

To perform the operations 10 e 11 correctly, it must obtain the angle of the voltages ($\theta = wt$). The main methods of obtaining the angle of the mains voltage are: method of zero crossing detection, filtering of the network voltages and PLL techniques. However, the use of a PLL is the most widely used technique due to its greater precision and less influence to the presence of harmonics and power grid disturbances in the grid [29].

There are various structures as described in [33]. However, the basic idea of operation of the PLL is to detect a difference between the instantaneous internal reference signal and external signal, as illustrated in Figure 14. The filter produces an error voltage proportional to the phase/frequency between the signals and operates in the VCO (Voltage Controlled Oscillator), which is a voltage controlled oscillator to change the frequency. So that internal matching the frequency of the signal external now obtains the angle of the external signal.

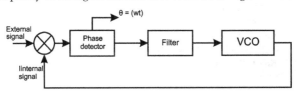

Figure 14. Basic idea of operation of the PLL.

The technique used in this work is known as three-phase SRF PPL (synchronous reference frame). This technique is the most widely used because it has little influence on the presence of harmonics and disturbances in the grid [34, 35].

Figure 15 shows the block diagram of three-phase PLL SRF. The basic operation of this PLL is to synchronize the synchronous reference frame PLL with the vector of the mains voltage.

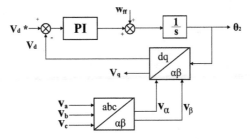

Figure 15. PPL synchronous reference frame

The voltages of the network (v_a, v_b, V_c) are obtained and then are transformed to stationary reference frame, using Equation 12. So has v_α e v_β using the proper angle estimated (θ_2), the variables v_α and v_β are transformed into the synchronous reference frame (Equation 13) resulting in tensionsv_d e v_q. Reference voltage v_d^* is set to zero. The error between v_d and v_d^* is processed by a PI controller that changes the value of (θ_2) in order to reset this error. When (θ_2) tends to the value of (θ_1) the sine tends to zero and the PLL is locked. In this situation, the value V_q is equal to the amplitude of the input voltages. The frequency obtained directly (w_{ff}) is added to improve the performance of the PLL.

$$\begin{bmatrix} v_\alpha \\ v_\beta \end{bmatrix} = \begin{bmatrix} 1 & -\frac{1}{2} & -\frac{1}{2} \\ 0 & \frac{\sqrt{3}}{2} & -\frac{\sqrt{3}}{2} \end{bmatrix} \begin{bmatrix} v_a(t) = V\cos(\theta_1) \\ v_b(t) = V\cos(\theta_1 - \frac{4\pi}{3}) \\ v_c(t) = V\cos(\theta_1 - \frac{2\pi}{3}) \end{bmatrix} = \begin{bmatrix} V\cos\theta_1 \\ -V sen\theta_1 \end{bmatrix} \tag{12}$$

$$\begin{bmatrix} v_d \\ v_q \end{bmatrix} = \begin{bmatrix} \cos\theta & sen\theta \\ -sen\theta & \cos\theta \end{bmatrix} \begin{bmatrix} v_\alpha \\ v_\beta \end{bmatrix} = \begin{bmatrix} -Vsen(\theta_1 - \theta_2) \\ V\cos(\theta_1 - \theta_2) \end{bmatrix} \tag{13}$$

5.6. SRM nonlinear model

Below will be described the operation of the model proposed by [36] that was developed for simulation in Matlab-Simulink software. This model is based on of magnetization curve that can be obtained by experiments, calculated by finite element or determined analytically by means of machine parameters that are available. The inputs of this model are the stator voltages in the phases of the machine and the outputs are the mechanical variables (torque, speed and rotor position).

5.6.1. General configuration model

The general configuration of the nonlinear simulation model can be seen in Figure 16. This model can be divided into 3 parts: model circuit, calculating the electromechanical torque and mechanical model.

The data of the magnetization curves of the machine are used to calculate the necessary magnetic characteristics in the model of the electrical circuit and for calculating the torque of the electromechanical machine.

5.6.2. Modeling the electrical circuit

The electrical circuit of a SRM of F stages, consisting of a resistor in series with an inductance is not linear for each phase of the machine. It has been that the flow equation for a phase j of MRV is given by:

$$\phi_j(t) = \int_0^t (V_j - R_j i_j) dt \tag{14}$$

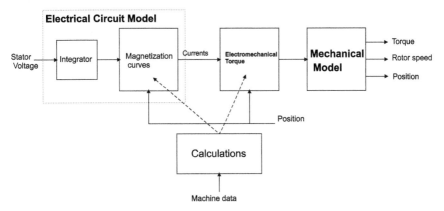

Figure 16. General configuration of the nonlinear model.

The currents in the stator phases are nonlinear functions $I(\phi, \theta)$ tha can be calculated from the magnetization curves $\phi(I, \theta)$. Therefore the electric circuit of a phase of MRV is modeled as shown in Figure 17.

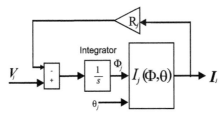

Figure 17. Electrical circuit model.

5.6.3. Magnetization curves.

The magnetization curves $\phi(I, \theta)$ are highly nonlinear due to the fact that MRV mainly operates in the saturation region. These curves can be obtained mainly in three ways: calculated by finite elements, analytical approach and experimental measurements.

5.6.4. Experimental measurements

The magnetization curves of MRV can be obtained through different forms as described in [37]. One way of measuring the magnetization curves are based on Equation 14. For each position of the rotor, a voltage is applied to the winding machine and the phase of the current and voltage are measured and stored. Then the magnetization curves are obtained from the processing of the waveforms of voltage and current. Figure 18 show the curves is observed in a test performed on an experimental MRV 8/6 obtained in [38].

This form of the curves is more accurate, but need experimental setup to perform measurements on the machine.

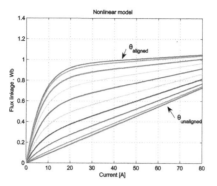

Figure 18. Experimental curver measurements.

5.6.5. Electromechanical torque

The electromechanical torque in SRM is the sum of the individual torques developed at each stage. When the magnetization curves are obtained experimentally or finite element electromechanical torque may be calculated using the equation of torque:

$$T_e = \frac{\delta W'_f}{\delta \theta} = \frac{\delta W'_f(i,\theta)}{\delta \theta} = \frac{dL(i,\theta)}{d\theta}\frac{i^2}{2} \tag{15}$$

Where (W'_f) is the coenergy.

5.6.6. Mechanical model

The equation of the mechanical model is given by:

$$T_m = T_{emag} - Dw - J\frac{dw}{dt} \tag{16}$$

The mechanical model is then modeled as shown in Figure 19.

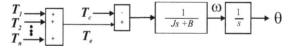

Figure 19. Mechanical model.

5.7. Machine model in Simulink

The nonlinear model developed by [36] became a block from the Simulink library SimPowerSystem. This block has an operating structure that can be viewed in Figure 20. From the data of magnetization of the two tables are created MRV. The table of current (ITBL) used in the model circuit and the torque table (TTBL) that gets the touch electromechanical each phase. For input values that do not exist in the tables the outputs are obtained by linear interpolation. The total electromechanical torque is obtained by summing the phases torques and sent to the mechanical model. The rotor position is obtained by integrating the speed of the machine. The bold lines refer to multiple streams of data that depends on the number of phases of the machine. The currents obtained from the table of the currents are generated at the terminals of the machine model.

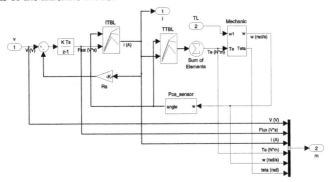

Figure 20. Diagram SRM model in Simulink.

6. Simulation results

The power control system proposed for the SRG connected to the grid was simulated using the Matlab-Simulink software . It was simulated a power profile (Figura 21) to be generated by the SRG in variable speed operation (Figura 23) and it observed that the reference active power was followed by the proposed DPC. In Figure 22 is observed the profile for the power factor of the energy sent to the grid and the output reach the reference..

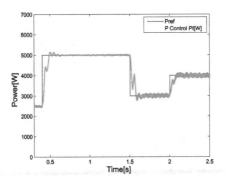

Figure 21. Power generated by SRG and the reference power.

Figure 24 shows the inverter phase currents of the SRG, it which observe the variation of current amplitudes, a fact that is justified due to the fact the controller changes θ_{off}. As can be seen in Figure 10. The voltage phases was set to 280 V and this should be the voltage of the link current which is controlled by the VSC.

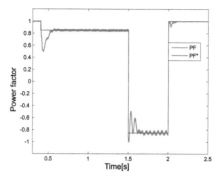

Figure 22. Power factor of the energy sent to the grid.

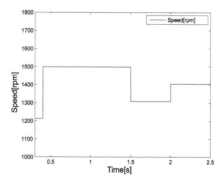

Figure 23. SRG operation speed.

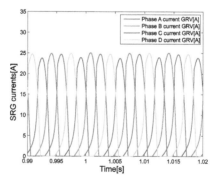

Figure 24. SRG current phases.

Figures 25 and 26 presents V_{dc} voltage and the phase a voltage and current for the operation of the SRG and allow to observe the performance of the control performed on the converter.

In Figure 26 is observed the voltage and phase current from the power grid. The THD (Total Harmonic Distortion) of the current sent to the grid analyzed by FFT (emph Fast Fourier Transform) (Figure 27) was 1.57%.

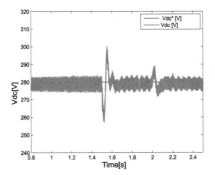

Figure 25. Voltage V_{dc} controlled by VSC.

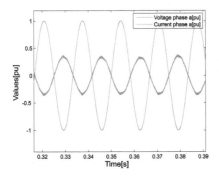

Figure 26. Phase a voltage and current grid.

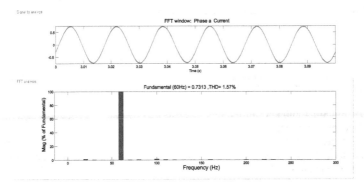

Figure 27. THD and harmonic components of the phase current.

7. Conclusions

This book chapter was presented a proposal for direct control of power for a SRG. The DPC controller allows the power control of the generator directly without using current loops. The controller has a satisfactory performance and no complexity for its implementation. The AHB converter allows robustness and the stages of regeneration energy and freewheel when needed. The simulation results confirm the effectiveness of the power controller during conditions of generator operation at variable speed and with different reference values of active power and power factor. Thus, the strategy of direct control of power is an interesting tool to control the power of the variable reluctance generator powered wind turbines.

Acknowledgements

The authors acknowledge FAPESP, CNPq and CAPES by the financial support.

Author details

Tárcio A. dos S. Barros[1,*],
Alfeu J. Sguarezi Filho[2] and Ernesto Ruppert Filho[1]

* Address all correspondence to: tarcio@dsce.fee.unicamp.br

1 Universidade Estadual de Campinas-UNICAMP, Faculdade de Engenharia Elétrica e Computação-FEEC, DSCE, Campinas, Brazil
2 Universidade Federal do ABC-UFABC, Santo André, Brazil

References

[1] Association World Wind Energy. *The World Wind Energy 2011 report*, 2012.

[2] McSwiggan D., L. Xu, and T Littler. Modelling and control of a variable-speed switched reluctance generator based wind turbine. *Universities Power Engineering Conference*, pages 459 – 463, June 2007.

[3] K. Ogawa, N. Yamamura, and M. Ishda. Study for small size wind power generating system using switched reluctance generator. *IEEE International Conference on Industrial Technology*, pages 1510–1515, 2006.

[4] R. Cardenas, R. Pena, M. Perez, G. Claro, J. Asher, and P. Wheeler. Control of a switched reluctance generator for variable-speed wind energy applications. *IEEE Transactions on energy conversrion*, 20(4):691 –703, December 2005.

[5] Alfeu J. Sguarezi Filho. Controle de potências ativa e reativa de geradores de indução trifásicos de rotor bobinado para aplicação em geração eólica com a utilização de controladores baseados no modelo matemático dinâmico do gerador. Tese doutorado, Faculdade de Engenharia Elétrica e Computação, Unicamp - Universidade Estadual de Campinas, Novembro 2010.

[6] Y. He, J. Hu, and Z. Rend. Modelling and control of wind-turbine used dfig under network fault condition. *Proceedings of the Eighth International Conf. on Electrical Machines and Systems*, 2:096–991, September. 2008.

[7] Seul-Ki Kim and Eung Kim. Pscad/emtdc-based modeling and analysis of a gearless variable speed wind turbine. *IEEE Transactions on energy conversrion*, 22(2):096–991, June 2007.

[8] Y. Chang and M. Liaw. Establisment of a switched reluctance generator-based common dc microgrid system. *IEEE transactions on power electronics*, pages 2512–2526, September. 2011.

[9] D.A. Torrey. Switched reluctance generators and their control. *Industrial Electronics, IEEE Transactions on*, 49(1):3 –14, feb 2002.

[10] T. Sawata. *The switched reluctance generator, Electronic Control of Switched Reluctance Machines*. Newness Power Engineering Series, 2001.

[11] S.F. Azongha, S. Balathandayuthapani, C.S. Edrington, and J.P. Leonard. Grid integration studies of a switched reluctance generator for future hardware-in-the-loop experiments. *Universities Power Engineering Conference*, pages 459 – 463, June 2010.

[12] Zhenguo Li, Jian Ma, Chunjiang Zhang, Dong-Hee Lee, and Jin-Woo Ahn. Research of switched reluctance wind power generator system based on variable generation voltage converter. In *Electrical Machines and Systems (ICEMS), 2010 International Conference on*, pages 418 –421, oct. 2010.

[13] H. Chen. Implementation of a three-phase switched reluctance generator system for wind power application. *IEEE International Conference on Industrial Technology*, (8):1 –6, June 2008.

[14] Baiming ShaD and Ali Emadi. A digital control for switched reluctance generators. *IEEE International Conference Mechatronics*, (4):182 –187, April 2000.

[15] Yilmaz Sozer and David A. Torrey. Closed loop control of excitation parameters for high speed switched-reluctance generators. *IEEE International Conference on Industrial Technology*, (4):1 –6, June 2000.

[16] K. Iordanis and C. Mademlis. Optimal efficiency control of switched reluctance generators. *IEEE Transactions on power electronics*, 21(4):1062–1071, April 2006.

[17] M. Godoy Simões and Felix A. Farret. *Renewable Energy Systems with Induction Generators*. CRC PRESS, 2004.

[18] Maurício B. C. Salles. Modelagem e análises de geradores eólicos de velocidade variável conectados em sistemas de energia elétrica. Tese doutorado, Escola Politécnica da Universidade de São Paulo, 2009.

[19] Manfred Stiebler. *Wind Energy Systems for Electric Power Generation*. Springer, 2008.

[20] R. Krishnan. *Switched Reluctance Motor Drives,Modeling, Simulation, Analysis, Design, and Applications.* CRC PRESS, 2001.

[21] Augusto Wohlgemuth Fleury Veloso da Silveira. Controle de tensão na carga para motor/gerador a relutância variável de três fases. Tese, Faculdade de Engenharia Elétrica, UFU, Universidade Federal de Uberlândia, Fevereiro 2011.

[22] E.A.E. Jebaseeli and D. Susitra. Performance analysis of various configurations of switched reluctance machine for wind energy applications. In *Recent Advances in Space Technology Services and Climate Change (RSTSCC), 2010,* pages 419 –423, nov. 2010.

[23] Gang Yang, Zhiquan Deng, Xin Cao, and Xiaolin Wang. Optimal winding arrangements of a bearingless switched reluctance motor. *Power Electronics, IEEE Transactions on,* 23(6):3056 –3066, nov. 2008.

[24] A. Takahashi, H. Goto, K. Nakamura, T. Watanabe, and O. Ichinokura. Characteristics of 8/6 switched reluctance generator excited by suppression resistor converter. *Magnetics, IEEE Transactions on,* 42(10):3458 –3460, oct. 2006.

[25] S.M. Muyeen. *Wind Energy Conversion Systems: Technology and Trends.* Green Energy and Technology. Springer, 2012.

[26] R. Cardenas, R. Pena, M.Perez, G. Asher, J. Clare, and P.Wheeler. Control system for grid generation of a switched reluctance generator driven by a variable speed wind turbine. *30th IEEE Industrial Electronics Society Conference,* pages 2–6, June 2004.

[27] K. Ogata. *Engenharia de Controle Moderno.* LTC, 2000.

[28] M. Kazmierkowski and L. Malesani. Current control techniques for three-phase voltage-source pwm converters: a survey. *Industrial Electronics, IEEE Transactions on Industrial Electronics,* 45(5):691 –703, October 1998.

[29] Marco Liserre Adrian V. Timbus Frede Blaabjerg, Remus Teodorescu. Overview of control and grid synchonization for distributed power generation systems. *IEEE Transactions on industrial electronics,* 53(5):691 –703, October 2008.

[30] M.A. Boost and P.D. Ziogas. State-of-the-art carrier pwm techniques: a critical evaluation. *Industry Applications, IEEE Transactions on,* 24(2):271 –280, mar/apr 1988.

[31] José R. Rodríguez, Juan W. Dixon, José R. Espinoza, Jorge Pontt, and Pablo Lezana. Pwm regenerative rectifiers: State of the art. *IEEE Transactions on Industrial Electronics,* 52(1), February 2005.

[32] José Antenor Polímio. *Eletrônica de potência para geração,transmissão e distribuição de energia,* 2012.

[33] R. M. Santos Filho, P. F. Seixas, P. C. Cortizo, L. A. B. Torres, and A. F. Souza. Comparison of three single-phase pll algorithms for ups applications. *Industrial Electronics, IEEE Transactions on,* 55(8):2923 –2932, aug. 2008.

[34] F. Liccardo, P. Marino, and G. Raimondo. Robust and fast three-phase pll tracking system. *Industrial Electronics, IEEE Transactions on*, 58(1):221 –231, jan. 2011.

[35] D. Jovcic. Phase locked loop system for facts. *Power Systems, IEEE Transactions on*, 18(3):1116 – 1124, aug. 2003.

[36] H. Le-Huy and P. Brunelle. A versatile nonlinear switched reluctance motor model in simulink using realistic and analytical magnetization characteristics. In *Industrial Electronics Society, 2005. IECON 2005. 31st Annual Conference of IEEE*, page 6 pp., nov. 2005.

[37] Wen Ding and Deliang Liang. A fast analytical model for an integrated switched reluctance starter/generator. *Energy Conversion, IEEE Transactions on*, 25(4):948 –956, dec. 2010.

[38] T.J.E. Miller and M. McGilp. Nonlinear theory of the switched reluctance motor for rapid computer-aided design. *Electric Power Applications, IEE Proceedings B*, 137(6):337 –347, nov 1990.

Chapter 3

Wireless Coded Deadbeat Power Control for Wind Energy Generation

C. E. Capovilla, A. J. Sguarezi Filho, I. R. S. Casella and
E. Ruppert

Additional information is available at the end of the chapter

1. Introduction

The growing demand for energy by the developing and developed countries, the search for energy alternatives to the use of fossil fuel and the recent special attention given to the environment, makes the study of alternative and renewable generation sources of electric energy in power grids extremely important [1].

Recently, the renewable power grids that carry electricity generated by wind, solar, and tidal have received new investments to turn out to be feasible and to optimize their use based on the concept of smart grids [2]. Among all sources of electric energy applied to this new concept, wind generation has emerged as one of the most promising presented techniques and has been the focus of several recent scientific studies [3, 4].

For a successful implementation, it is necessary to develop a complete telecommunications framework composed by communication networks, data management, and real-time monitoring applications with a strong interaction. In particular, the application of a modern telecommunication system for controlling and monitoring in smart grids applications requires a complex infrastructure for an efficient operation [5], and its development and operability presents several non-trivial issues due to the convergence of different areas of knowledge and design aspects.

In this way, wireless communications appear as an interesting solution for offering many benefits such as low cost of development, expansion facilities, possibility of using the technologies currently applied in mobile telephone systems, flexibility of use, and distributed management. However, wireless transmissions are subject to distortions and errors caused by the propagation channel that can cause serious problems to the controlled and monitored equipments, thus, to the energy plant as a whole. This intrinsic problem of wireless communication systems can be circumvented through the use of Forward Error Correction

(FEC) [6]. This coding technique is used in all modern wireless digital systems and is essential to ensure the integrity of information, reducing significantly the Bit Error Rate (BER) and the latency of the information by adding redundancy to the transmitted information [7].

There are currently several different schemes of FEC that are used in commercial wireless communication systems, for instance, the Reed Solomon (RS) coding [8], Convolutional Coding (CONV) [9], Turbo Coding (TC) [10, 11], and Low Density Parity Check (LDPC) coding [12–14]. Among them, the LDPC coding is the one that presents the best performance, approaching significantly to the limits set by the seminal work of Shannon [15] and that shows an excellent compromise between decoding complexity and performance [13, 16]. Besides, the LDPC coding has recently been added to the IEEE 802.16e Standard, commonly known as Worldwide Interoperability for Microwave Access (WiMAX) for mobile applications [17].

It is worth noting that there are some works in the scientific literature referencing the application of wireless technology for monitoring wind energy systems based on sensor networks [18, 19], however there has not been presented yet any deep research about the use of wireless technology for control applications in these systems, making it difficult to estimate the real impact of its use or its advantages and difficulties.

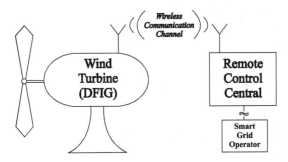

Figure 1. Wireless System Control Schematic.

Concerning the generators used in wind turbines, the Doubly-Fed Induction Generators (DFIG) have been widely used in these systems [20] due to their great characteristics. The main advantages of using DFIG are their ability to operate at variable speeds and control the active and reactive power into four quadrants, in contrast, for instance, to the Squirrel Cage Induction Generators (SCIG), which operate at fixed speed [1, 21]. The active and reactive power control of DFIG is made by using field orientation. In the work [22], some investigations were carried out to power control of a DFIG using Proportional-Integral (PI) controller, however this type of controller has problems related to the design of their gain due to operating conditions of the generator. In the works [23], [24], and [25] other investigations were done, respectively, for the use of predictive control techniques and internal mode control. Although both controllers show a satisfactory performance, they present many difficulties to implementation due to their intrinsic formulations.

In this context, this work proposes a wireless coding deadbeat power control using DFIG machine to improve robustness and reliability of the generation system, as shown in Fig. 1

(simplified schematic). The proposed controller is based on discrete dynamic mathematical model of the generator and it uses the vector control technique that allows the independent control of active and reactive power.

The wireless communication system is used to send the power reference signals to the DFIG controller applying an LDPC coding scheme to reduce the transmission errors and the overall system latency. The performance of the proposed system is investigated for different radio propagation scenarios to evaluate the real impact of the wireless transmission in the wind energy control system. It is noteworthy that the errors generated in the wireless transmission cannot be easily removed without using advanced FEC coding techniques similar to those presented in this work.

Although the proposed system has been analyzed for a single link between one remote control unit and one aerogenerator, it can easily be expanded to control several aerogenerators for wind farm applications. The use of wireless communication in wind farms becomes very interesting for technical and economic reasons. The work of [26] shows a wireless remote control for a wind farm consisting of offshore wind generation platforms. The choice of an appropriate control system and a wireless monitoring becomes essential for this type of application, due to its easy deployment avoiding the need for submarine optical fibers that have high cost of installation and maintenance. Besides, any changes in the offshore platforms positioning due to climate or hydrological characteristics would not be problem for the wireless communication system. The importance of that the communication systems have on the effective control and maintenance of wind farms is discussed in the work of [27]. A brief description of IEC61400-25 "Communications for Monitoring and Control of Wind Power Plants" is provided. As a case study, it was analyzed the Horns Rev offshore wind farm in Denmark that employs a principal communication system based on optical fiber and a secondary wireless system, both integrated into the Supervisory Control And Data Acquisition (SCADA) system, linking wind turbines to the onshore control center. Additionally, the work of [28] has shown some problems present in a real wired system to control and monitor wind turbines based on Lonworks and the authors present as solution a wireless control and monitoring system that offers many facility and benefits.

These works bring evidences and exemplify the actual advantages and features offered by the use of wireless communications, but none of them proposes or examines techniques that can ensure the reliability and security for control and monitoring information on transmission error robustness, due to the degrading effects of wireless communication channel. Thus, this work aims to fill a gap in the literature to demonstrate the functional viability of the use of wireless systems for this type of application when an appropriate coding technique is applied.

The chapter is organized as follows: DFIG adaptive deadbeat power control is shown in section 2, the wireless coding communication is presented in section 3, main results are considered in section 4, and section 5 concludes the chapter.

2. DFIG Deadbeat power control

The doubly-fed induction machine in synchronous reference frame can be represented [29] by:

$$\vec{v}_{1dq} = R_1 \vec{i}_{1dq} + \frac{d\vec{\lambda}_{1dq}}{dt} + j\omega_1 \vec{\lambda}_{1dq} \tag{1}$$

$$\vec{v}_{2dq} = R_2 \vec{i}_{2dq} + \frac{d\vec{\lambda}_{2dq}}{dt} + j\left(\omega_1 - B\omega_r\right)\vec{\lambda}_{2dq} \tag{2}$$

where: $\vec{v}_{dq}, \vec{i}_{dq}, \vec{\lambda}_{dq}$ are, respectively, voltage, current, and flux space vectors in synchronous reference frame dq, R is the resistance of the winding, L is the inductance of the winding, B is the number of pairs of poles, ω_1 is the synchronous speed, ω_r is the rotor speed, and the subscripts 1 and 2 denote, respectively, stator and rotor parameters.

The generator active and reactive power can be obtained by:

$$P = \frac{3}{2}\left(v_{1d}i_{1d} + v_{1q}i_{1q}\right) \tag{3}$$

$$Q = \frac{3}{2}\left(v_{1q}i_{1d} - v_{1d}i_{1q}\right) \tag{4}$$

Using stator flux oriented control, that decouples d and q components, the relationship between stator and rotor currents becomes:

$$i_{1d} = \frac{\lambda_1}{L_1} - \frac{L_M}{L_1}i_{2d} \tag{5}$$

$$i_{1q} = -\frac{L_M}{L_1}i_{2q} \tag{6}$$

where: v, i, λ are, respectively, voltage, current and flux magnitudes and L_M is the mutual inductance.

The active (3) and reactive (4) power can be calculated by using (5) and (6) and it is given by:

$$P = -\frac{3}{2}v_1\frac{L_M}{L_1}i_{2q} \tag{7}$$

$$Q = \frac{3}{2}v_1\left(\frac{\lambda_1}{L_1} - \frac{L_M}{L_1}i_{2d}\right) \tag{8}$$

Equations (8) and (7) show the system can provide independent active and reactive power control by regulating the rotor current. The proposed deadbeat power control, shown in

Fig. 2, considers these relationships. Consequently, stator active and reactive power control can be accomplished by using rotor current control of the DFIG with stator directly connected to the grid.

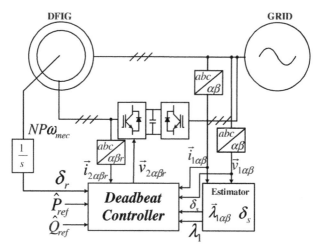

Figure 2. Deadbeat Power Control Block Diagram.

The discretized rotor equation (based on the zero-order hold method) in the synchronous referential frame, using equation (2), the stator flux position at sampling time $k+1$ and using equations (5) and (6), can be represented [30] by:

$$\begin{bmatrix} i_{2d}(k+1) \\ i_{2q}(k+1) \end{bmatrix} = \begin{bmatrix} 1 - \frac{R_2 T}{\sigma L_2} & \frac{\omega_{sl} T}{\sigma} \\ \frac{-\omega_{sl} T}{\sigma} & 1 - \frac{R_2 T}{\sigma L_2} \end{bmatrix} \begin{bmatrix} i_{2d}(k) \\ i_{2q}(k) \end{bmatrix} +$$

$$+ \begin{bmatrix} \frac{T}{\sigma L_2} & 0 \\ 0 & \frac{T}{\sigma L_2} \end{bmatrix} \begin{bmatrix} v_{2d}(k) \\ v_{2q}(k) \end{bmatrix} +$$

$$+ \begin{bmatrix} 0 & \frac{\omega_{sl} L_M T}{\sigma L_2} \\ \frac{-\omega_{sl} L_M T}{\sigma L_2} & 0 \end{bmatrix} \begin{bmatrix} i_{1d}(k) \\ i_{1q}(k) \end{bmatrix} \qquad (9)$$

where: $\omega_{sl} = \omega_1 - B\omega_r$ is the slip frequency and $\sigma = 1 - L_M^2/(L_1 L_2)$.

The rotor voltage which is calculated to guarantee null steady state error by using the deadbeat theory [31] is given by:

$$v_{2d}(k) = \sigma L_2 \frac{i_{2d}(k+1) - i_{2d}(k)}{T} + R_2 i_{2d}(k) +$$

$$- L_2 \omega_{sl} i_{2q}(k) - L_M \omega_{sl} i_{1q}(k) \qquad (10)$$

$$v_{2q}(k) = \sigma L_2 \frac{i_{2q}(k+1) - i_{2q}(k)}{T} + R_2 i_{2q}(k) +$$
$$+ L_2 \omega_{sl} i_{2d}(k) + L_M \omega_{sl} i_{1d}(k) \tag{11}$$

For the active power control, the rotor current reference by using (7) is given by:

$$i_{2q}(k+1) = i_{2q_{ref}} = -\frac{2P_{ref}L_1}{3v_1 L_M} \tag{12}$$

and for the reactive power control by using (8) is:

$$i_{2d}(k+1) = i_{2d_{ref}} = -\frac{2Q_{ref}L_1}{3v_1 L_M} + \frac{\lambda_1}{L_M} \tag{13}$$

where: P_{ref} is the active power reference and Q_{ref} is the reactive power reference.

Thus, if the d and q axis voltage components calculated according to equations (10), (11), (12), and (13) are applied to the generator then, the active and reactive power convergence to their respective commanded values will occur in one sampling interval. The desired rotor voltage in the rotor reference frame $(\delta_s - \delta_r)$ generates switching signals for the rotor side using either space vector modulation.

Stator currents and voltages, rotor speed and currents are measured to stator flux position δ_s and magnitude λ_1, synchronous frequency ω_1 and slip frequency ω_{sl} estimation.

2.1. Stator flux estimation

For a Deadbeat power control, as shown in the equations (10) and (11), it is required to calculate the active and reactive power values, their errors, the stator flux magnitude and position, the slip speed and synchronous frequency.

The stator flux $\vec{\lambda}_{1\alpha\beta}$ estimation in stationary reference frame is given by:

$$\vec{\lambda}_{1\alpha\beta} = \int \left(\vec{v}_{1\alpha\beta} - R_1 \vec{i}_{1\alpha\beta} \right) dt \tag{14}$$

This expression can be implemented to allow the estimation of the stator flux, even though the induction machine is operating at low speed in a direct torque control system, as shown in [32].

Thus, the stator flux position by using equation (14) is given by:

$$\delta_s = \arctan\left(\frac{\lambda_{1\beta}}{\lambda_{1\alpha}}\right) \tag{15}$$

The synchronous speed ω_1 estimation is:

$$\omega_1 = \frac{d\delta_s}{dt} = \frac{\left(v_{1\beta} - R_1 i_{1\beta}\right)\lambda_{1\alpha} - \left(v_{1\alpha} - R_1 i_{1\alpha}\right)\lambda_{1\beta}}{(\lambda_{1\alpha})^2 + (\lambda_{1\beta})^2} \tag{16}$$

and the slip speed estimation by using the rotor speed and synchronous speed is:

$$\omega_{sl} = \omega_1 - B\omega_{mec} \tag{17}$$

The angle between stator and rotor flux is given by:

$$\delta_s - \delta_r = \int \omega_{sl}dt \tag{18}$$

3. Wireless coding communication

The proposed wireless control system, shown in Fig. 3, uses LDPC codes [9, 12, 33] to improve system performance and reliability.

The LDPC are (N_c, N_b) binary linear block codes that have a sparse parity-check matrix **H** that can be described in terms of a Tanner graph [34], where each bit in the codeword corresponds to a variable node and each parity-check equation corresponds to a check node. A check node j is connected to a variable node k whenever the element $h_{j,k}$ in **H** is equal to 1 [9, 34].

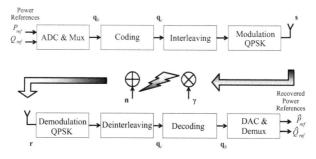

Figure 3. Wireless Coding Communication Diagram.

Extended Irregular Repeat Accumulate (eIRA) codes [14, 35–38] are a special subclass of LDPC codes that improve the systematic encoding process and generate good irregular LDPC codes for high code rate applications. The eIRA parity-check matrix can be represented by $\mathbf{H} = [\mathbf{H_1} \ \mathbf{H_2}]$, where $\mathbf{H_1}$ is a sparse (N_m) by (N_c) matrix, that can be constructed irregularly by density evolution according to optimal weight distribution [14], and $\mathbf{H_2}$ is the (N_m) by (N_m) dual-diagonal square matrix given by:

$$\mathbf{H_2} = \begin{bmatrix} 1 & & & & \\ 1 & 1 & & & \\ & 1 & \ddots & & \\ & & \ddots & 1 & \\ & & & 1 & 1 \end{bmatrix} \tag{19}$$

where: N_b is the number of control bits, N_c is the number of coded bits and N_m is the number of parity bits.

Given the constraint imposed on the $\mathbf{H}$ matrix, the generator matrix can be represented in the systematic form by the (N_b) by (N_c) matrix:

$$\mathbf{G} = [\mathbf{I} \ \mathbf{\Psi}] \tag{20}$$

where: $\mathbf{I}$ is the identity matrix, $\mathbf{\Psi} = \mathbf{H_1^T} \times \mathbf{H_2^{-T}}$ and $\mathbf{H_2^{-T}}$ is the upper triangular matrix given by:

$$\mathbf{H_2^{-T}} = \begin{bmatrix} 1 & 1 & 1 & \cdots & 1 & 1 \\ & 1 & 1 & \cdots & 1 & 1 \\ & & 1 & & 1 & 1 \\ & & & \ddots & \vdots & \vdots \\ & & & & 1 & 1 \\ & & & & & 1 \end{bmatrix} \tag{21}$$

The encoding process can be accomplished by first multiplying the control information vector $\mathbf{q_b} = \begin{bmatrix} q_b(1) \cdots q_b(N_b) \end{bmatrix}^T$ by the sparse matrix $\mathbf{H_1^T}$ and then differentially encoding this partial result to obtain the parity bits. The systematic codeword vector $\mathbf{q_c} = \begin{bmatrix} q_c(1) \cdots q_c(N_c) \end{bmatrix}^T$ can be simply obtained by combining the control information and the parity bits:

$$\mathbf{q_c} = [\mathbf{q_b} \ \mathbf{\Psi}] \tag{22}$$

In the transmission process, the codeword vector is then interleaved and Quaternary Phase Shift Keying (QPSK) mapped using Gray code [7], resulting in the symbol vector $\mathbf{s} = \left[s\left(1 \right) \cdots s\left(N_s \right) \right]^T$, where N_s is the number of transmitted coded control symbols. Afterwards, the coded symbols are filtered, upconverted and transmitted by the wireless fading channel.

Assuming that the channel variations are slow enough that intersymbol interferences (ISI) can be neglect, the fading channel can be modeled as a sequence of zero-mean complex Gaussian random variables with autocorrelation function [7, 39]:

$$R_h\left(\tau \right) = J_0\left(2\pi f_D T_s \right) \tag{23}$$

where: $J_0()$ is the zero-th order Bessel function, T_s is the signaling time and f_D is the Doppler spread.

Thus, in the receive process, the complex low-pass equivalent discrete-time received signal can be represented by [7]:

$$\mathbf{r} = \gamma \cdot \mathbf{s} + \mathbf{n} \tag{24}$$

where: $\mathbf{r} = \left[r\left(1 \right) \cdots r\left(N_s \right) \right]^T$ is the received signal vector, $\gamma = \left[\gamma\left(1 \right) \cdots \gamma\left(N_s \right) \right]^T$ is the vector of complex coefficients of the channel and $\mathbf{n} = \left[n\left(1 \right) \cdots n\left(N_s \right) \right]^T$ is the Additive White Gaussian Noise (AWGN) vector. Note that the above vector multiplication is performed element by element.

Once the transmitted vector $\mathbf{s}$ is estimated, considering perfect channel estimation, the transmitted control bits can be recovered by performing symbol demapping, code deinterleaving and bit decoding.

Decoding can be accomplished by a message passing algorithm [16, 40–42] based on the Maximum A Posteriori (MAP) criterion [9], that exchanges soft-information iteratively between the variable and check nodes. The exchanged messages can be represented by the following Log-Likelihood Ratio (LLR):

$$L_{c_k} = \log \left[\frac{p(\mathbf{q}_c\left(k \right) = 0|\mathbf{d})}{p(\mathbf{q}_c\left(k \right) = 1|\mathbf{d})} \right] \tag{25}$$

The LLR message from the jth check node to the kth variable node is given by:

$$L_{r_{j,k}} = 2\,\text{atanh} \left[\prod_{k' \in V_{j\backslash k}} \tanh \left(\frac{L_{q_{k',j}}}{2} \right) \right] \tag{26}$$

The set V_j contains the variable nodes connected to the jth check node and the set C_k contains the check nodes connected to the kth variable node. $V_{j\setminus k}$ is the set V_j without the kth element, and $C_{k\setminus j}$ is the set C_k without the jth element. The LLR message from the kth variable node to the jth check node is obtained by:

$$L_{q_{k,j}} = L_{c_k} + \sum_{j' \in c_{k\setminus j}} L_{r_{j',k}} \tag{27}$$

and the LLR for the kth code bit is given by:

$$L_{Q_k} = L_{c_k} + \sum_{j \in c_k} L_{r_{j,k}} \tag{28}$$

At the end of each iteration, L_{Q_k} provides an updated estimate of the a posteriori LLR of the transmitted coded bit $q_c(k)$. If $L_{Q_k} > 0$, then $q_c(k) = 1$, else $q_c(k) = 0$.

4. Control system performance

The deadbeat power control strategy, for this analysis, has a sampling time of 0.5×10^{-4}s and the DFIG parameters are shown in Appendix. During the period of 1.75-2.0s, the rotor speed was increased from 151 to 226.5 rad/s to include also the wind variation in the analysis. In the simulations, the active and reactive power references were step changed, respectively, from -100 to -120 kW and from 60 to 0 kvar at 1.25s. At 1.5s, the references also were step changed from -120 to -60 kW and from 0 to -40 kvar. Again, at 1.75s, the references were step changed from -60 to -100 kW and from -40 to -60 kvar. These references are the inputs of the wireless coding power control, shown in Fig. 3, which is analyzed for two different scenarios: an AWGN channel and a more realistic flat fading correlated Rayleigh channel.

The system is evaluated for a frequency flat fading Rayleigh channel with a Doppler spread of 180 Hz. The LDPC coding scheme uses the (64,800; 32,400) eIRA code specified in [43], and an ordinary Convolution Coding scheme with a (171, 133) generator polynomial with constrain length of 7 is used as reference of performance [9]. Both schemes have code rate of 1/2 and employ a random interleaving of length 64,800. For simplicity, the number of iterations in the LDPC decoding is limited to 25. The bit duration is 1.25×10^{-5}s and each transmitted frame is composed by 32,400 QPSK coded symbols. In Fig. 4 and 5, the final responses of the wireless power control system employing CONV are presented for a typical Noise to Signal Ratio (E_b/N_0) of 10 dB. The spikes presented in the responses of the system occur due to the errors in the wireless communication, even with the use of a very efficient error correction scheme. It can be observed that several of these spikes, presented in the reference signals, are followed by the controller and not by others due to the fact that the time response of the controller is not sufficient to follow quick changes caused by destructive effects of the channel in the transmitted signal.

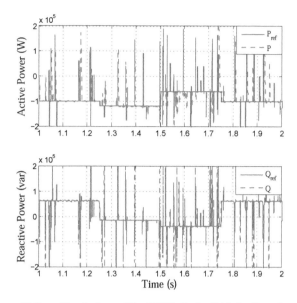

Figure 4. Step Response of Active and Reactive Powers Using CONV Coding in a Flat Fading Channel.

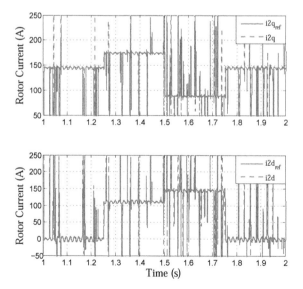

Figure 5. Step Response of Rotor Current $\vec{i}_{2_{dq}}$ Using CONV Coding in a Flat Fading Channel.

These errors in the control system can permanently damage the aerogenerator, the wind generation system, or even cause a loss of system efficiency, since the machine will not generate its maximum power track at that moment, and additionally, they generate undesirable harmonic components to the power grid. The damage related to wind generation occurs due to the fact that high values of $\frac{di}{dt}$, as shown in Fig. 6, can completely deteriorate the Insulated Gate Bipolar Transistors (IGBTs) and, consequently, through the power converter, can cause short circuits in rotor and/or stator of the generator.

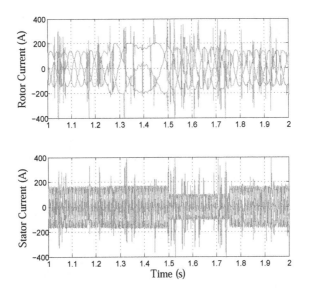

Figure 6. Stator and Rotor Currents Using CONV Coding in a Flat Fading Channel.

Thus, it is necessary to use a wireless control system capable of minimizing the occurrence of these spikes arising from errors caused by the channel distortions. Aiming this, it is highlighted the proposal of using a more robust wireless control system based on LDPC coding. Fig. 7 and 8 show the response of the wireless controller employing the LDPC coding scheme for the same E_b/N_0 of 10 dB and step reference signals described at the beginning of this section.

The satisfactory performance of the wireless control system can be seen due to the fact the references were perfectly followed by the controller and the inexistence of destructive spikes caused by errors in the wireless transmission system. Additionally, these good functionalities are shown in Fig. 9, where the stator currents present expected waveforms for an good operational functionality.

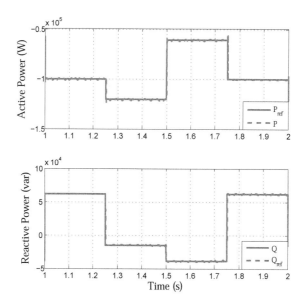

Figure 7. Step Response of Active and Reactive Powers Using LDPC Coding in a Flat Fading Channel.

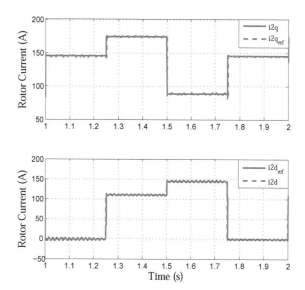

Figure 8. Step Response of Rotor Current $\vec{i}_{2_{dq}}$ Using LDPC Coding in a Flat Fading Channel.

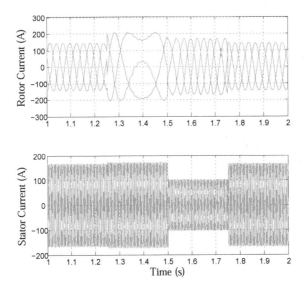

Figure 9. Stator and Rotor Currents Using LDPC Coding in a Flat Fading Channel.

To complete the analysis, it is evaluated the performance of the proposed wireless coded deadbeat power control system for many values of E_b/N_0 in a flat fading channel. In Fig. 10, a comparison of performance for No Coding, CONV, and LDPC schemes is presented. As expected, the performance of LDPC is significantly superior than CONV. As pointed out, the performance improvement of LDPC over CONV for a BER of 10^{-5} is approximately 26.8 dB, demonstrating the good performance of LDPC in this channel condition.

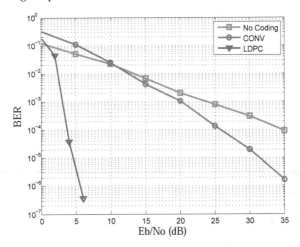

Figure 10. Performance Comparison for Different Coding Schemes in Flat Fading Channel (180 Hz - QPSK).

Table 1 shows the number and percentage of phase inversions, corresponding to P_{ref} and Q_{ref} references, presented in the recovered control signal for both scenarios. It can be seen that LDPC coding requires a significant lower E_b/N_0 to present the same order of phase inversions as CONV.

For a low BER as 10^{-5}, there are some changes in the active and reactive power references that can cause serious problems in the generator, and consequentially, in the energy plant. However, the use of LDPC coding can reduce notably this number for a typical E_b/N_0 in real systems and can improve considerably system robustness to the channel impairments. For instance, a system operating with an E_b/N_0 of 10 dB employing CONV will fail dramatically, while a system using LDPC coding will be free of communication errors, showing the real operational viability of the proposed control technique in wind power plants.

Coding Scheme	Bit Error Rate	Eb/No (dB)	Number of Inversions
CONV	10^{-3}	20.0	167 (0.087%)
CONV	10^{-4}	25.5	21 (0.01%)
CONV	10^{-5}	31.5	4 (0.002%)
LDPC	10^{-3}	3.20	188 (0.098%)
LDPC	10^{-4}	3.80	26 (0.013%)
LDPC	10^{-5}	4.70	1 (0.0005%)

Table 1. Control Inversion in a Flat Fading Channel.

5. Conclusion

This work proposes a wireless coding control system using a deadbeat controller applied to a doubly-fed induction aerogenerator for smart grid applications. An analysis for different coding schemes shows that, even for a relatively low BER, the power reference changes can occur and it can be very dangerous for the generator and the energy plant. However, the use of LDPC coding improves significantly the robustness of the system in severe noise and fading channel situations, eliminating the occurrence of errors in the active and reactive power references for operational conditions.

In addition, it is noteworthy that the errors generated in the wireless transmission cannot be easily removed without using advanced FEC coding techniques similar to the one proposed in this work without increasing significantly the quantity of retransmissions, that should be controlled, for instance, by an Automatic Repeat Request (ARQ) scheme implemented in an upper communication layer. On the other hand, depending on channel conditions, even employing an efficient ARQ scheme, the latency of the system can increase dramatically without a similar technique as the proposed in this paper. Alternatively, as a future work, other types of codification and modulation can be explored.

Appendix

Doubly-fed induction generator parameters are: [44]:
$R_1 = 24.75 \ m\Omega$; $R_2 = 13.3 \ m\Omega$; $L_M = 14.25 \ mH$; $L_{l1} = 284 \ \mu H$; $L_{l2} = 284 \ \mu H$; $J = 2.6 \ Kg \cdot m^2$;
$B = 2$; $P_N = 149.2 \ kVA$; $V_N = 575 \ V$.

where: R - Winding resistance, L_M - Mutual inductance, L_l - Dispersion inductance, J - Moment of inertia, B - Number of pair poles, P_N - Nominal Power, V_N - Nominal Voltage. The subscripts 1 and 2 represent the stator and rotor parameters, respectively.

Author details

C. E. Capovilla[1], A. J. Sguarezi Filho[1],
I. R. S. Casella[1] and E. Ruppert[2]

1 Universidade Federal do ABC - UFABC, Brazil
2 Universidade Estadual de Campinas - Unicamp, Brazil

References

[1] M. Godoy SimÃțes and Felix A. Farret. *Renewable Energy Systems with Induction Generators*. CRC Press, 2004.

[2] J. Blau. Europe plans a north sea grid. *IEEE Spectrum*, pages 08–09, March 2010.

[3] M. Glinkowski, J. Hou, and G. Rackliffe. Advances in wind energy technologies in the context of smart grid. *Proceedings of the IEEE*, 99(6):1083–1097, June 2011.

[4] J. Wang, X. Du, and X. Zhang. Comparison of wind power generation interconnection technology standards. *Asia-Pacific Power and Energy Engineering Conference*, March 2011.

[5] R. Strzelecki and G. Benysek. *Power Electronics in Smart Electrical Energy Networks*. Springer-Verlag, 2008.

[6] T. J. Li. *Low complexity capacity approaching schemes: Design, analysis, and applications*. Ph.D. dissertation, Texas AM Univ., 2002.

[7] J. G. Proakis. *Digital Communications*. MCGraw-Hill, 2001.

[8] J. Jiang and K. R. Narayanan. Iterative soft decision decoding of Reed Solomon. *IEEE Communications Letters*, 8:244–246, 2004.

[9] S. Lin and D. J. Costello. *Error control coding*. Prentice Hall, 2004.

[10] C. Berrou, A. Glavieux, and P. Thitimajshima. Near Shannon limit error-correcting coding and decoding: Turbo-codes. *IEEE International Communications Conference*, pages 1064–1070, 1993.

[11] J. Chen and A. Abedi. Distributed turbo coding and decoding for wireless sensor networks. *IEEE Communications Letters*, 15:166–168, 2011.

[12] R. G. Gallager. *Low-Density Parity-Check Codes*. Cambridge, 1963.

[13] D. J. C. MacKay and R. M. Neal. Near Shannon limit performance of low-density parity-check codes. *IET Electronics Letters*, 32:1645–1646, 1996.

[14] Y. Zhang, W. E. Ryan, and Y. Li. Structured eIRA codes with low floors. *Proceedings of the International Symposium on Information Theory*, pages 174–178, September 2005.

[15] E. C. Shannon. A mathematical theory of communication. *Bell System Technical Journal*, 27:379–423, 1948.

[16] T. Richardson, A. Shokrollahi, and R. Urbanke. Design of capacity-approaching low-density parity-check codes. *IEEE Transactions on Information Theory*, 47:619–637, February 2001.

[17] IEEE. Standard for local and metropolitan area networks, part 16: Air interface for fixed and mobile wireless access systems. *IEEE Std. 802.16-2004*, 2004.

[18] M. Adamowicz, R. Strzelecki, Z. Krzeminski, J. Szewczyk, and L. Lademan. Application of wireless communication to small WECS with induction generator. *IEEE Mediterranean Electrotechnical Conference*, pages 944–948, June 2010.

[19] M. Adamowicz, R. Strzelecki, Z. Krzeminski, J. Szewczyk, and L. Lademan. Wireless short-range device for wind generators. *12th Biennial Baltic Electronics Conference*, pages 1736–3705, November 2010.

[20] J. F. Manwell, J. G. McGowan, and A. L. Rogers. *Wind Energy Explained: Theory, Design and Application*. 2 edition, 2010.

[21] R. Datta and V. T. Rangathan. Variable-speed wind power generation using doubly fed wound rotor induction machine - A comparison with alternative schemes. *IEEE Trans. on Energy Conversion*, 17(3):414–421, September 2002.

[22] Arantxa Tapia, Gerardo Tapia, J. Xabier Ostolaza, and JosÃÍ RamÃşn SÃąenz. Modeling and control of a wind turbine driven doubly fed induction generator. *IEEE Trans. on Energy Conversion*, (194-204), June 2003.

[23] Zhang Xin-fang, XU Da-ping, and LIU Yi-bing. Predictive functional control of a doubly fed induction generator for variable speed wind turbines. *IEEE World Congress on Intelligent Control and Automation*, June 2004.

[24] J. Morren and S. W. H. Haan. Ridethrough of wind turbines with doubly-fed induction generator during a voltage dip. *IEEE Transactions on Energy Conversion*, 20(2):435–441, June 2005.

[25] Ji. Guo, X. Cai, and Y. Gong. Decoupled control of active and reactive power for a grid-connected doubly-fed induction generator. *Third International Conference on Electric Utility Deregulation and Restructuring and Power Technologies*, pages 2620 – 2625, April 2008.

[26] Z. Li, F. Zheng, Y. Wu, and H. Gao. Offshore wind farm construction platform jack up control system. *World Non-Grid-Connected Wind Power and Energy Conference*, pages 24–26, September 2009.

[27] O. Anaya-Lara, N. Jenkins, and J. R. McDonald. Communications requirements and technology for wind farm operation and maintenance. *IEEE International Conference on Industrial and Information Systems*, pages 173–178, August 2006.

[28] C. Wanzhi, T. Zhiyong, Z. Quangui, and C. Liang. Research of wireless communication based on lonworks for wind turbine control system. *IEEE International Conference on Energy and Environment Technology*, pages 787–789, October 2009.

[29] D. W. Novotny and T. A. Lipo. *Vector Control and Dynamics of AC Drives*. Clarendon Press Oxford, 1996.

[30] A. J. Sguarezi Filho and E. Ruppert. A deadbeat active and reactive power control for doubly-fed induction generators. *Electric Power Components and Systems*, 38(5):592–602, 2010.

[31] G. F. Franklin, J. D. Powel, and M. Workman. *Digital Control of Dynamic Systems*. Addison-Wesley Publishing Company, 1994.

[32] A. J. Sguarezi Filho and E. Ruppert. The complex controller for three-phase induction motor direct torque control. *SBA - Controle e automaÃğÃčo*, 20(2):256–262, 2009.

[33] Y. Zhang and W. E. Ryan. Toward low LDPC-code floors: a case stud. *IEEE Transactions on Communications*, 57(6):1566–1573, June 2009.

[34] R. M. Tanner. A recursive approach to low complexity codes. *IEEE Transactions on Information Theory*, 27(5):533–547, September 1981.

[35] H. Jin, A. Khandekar, and R. J. McEliece. Irregular repeat-accumulate codes. *Proc. Int. Symp. Turbo Codes and Related Topics*, pages 1–5, September 2000.

[36] M. Yang and W. E. Ryan. Lowering the error-rate floors of moderate length high-rate irregular LDPC codes. *Int. Symp. Information Theory*, 2:237, July 2003.

[37] M. Yang, W. E. Ryan, and Y. Li. Design of efficiently encodable moderate-length high-rate irregular LDPC codes. *IEEE Transactions on Communications*, 52(4):564–571, April 2004.

[38] J. Kim, A. Ramamoorthy, and S. Mclaughlin. The design of efficiently-encodable rate-compatible LDPC codes. *IEEE Transactions on Communications*, 57:365–375, 2009.

[39] A. Barbieri, A. Piemontese, and G. Colavolpe. On the ARMA approximation for frequency-flat rayleigh fading channels. *IEEE International Symposium on Information Theory*, pages 1211–1215, June 2007.

[40] T. Richardson and R. Urbanke. The capacity of low-density parity check codes under message-passing decoding. *IEEE Transactions on Information Theory*, 47:599–618, February 2001.

[41] L. Dinoi, F. Sottile, and S. Benedetto. Design of versatile eIRA codes for parallel decoders. *IEEE Transactions on Communications*, 56(12):2060–2070, 2008.

[42] B. Shuval and I. Sason. On the universality of LDPC code ensembles under belief propagation and ml decoding. *IEEE 26th Convention of Electrical and Electronics Engineers*, pages 355–359, 2010.

[43] ETSI. DVB-S.2. *Standard Specification*, pages 302–307, March 2005.

[44] A. J. Sguarezi Filho, M. E. de Oliveira Filho, and E. Ruppert. A predictive power control for wind energy. *IEEE Transactions on Sustainable Energy*, 2(1):97–105, January 2011.

Chapter 4

Wind Turbines Reactive Current Control During Unbalanced Voltage Dips

Ivan Jorge Gabe , Humberto Pinheiro and
Hilton Abílio Gründling

Additional information is available at the end of the chapter

1. Introduction

These chapter deals with the current references generation for the control of grid-connected voltage source converters, used in Wind Energy Generating Units (WEGU),operating under balanced and unbalanced voltage sags. The significant growing of power generation by wind turbines prompted grid operators to update the grid connection requirements. In the Early grid codes, WEGU disconnection were allowed during voltages dips and frequency disturbances. However, nowadays, they are required to stay connected during voltage dips resulted from faults at the grid side as well as support the grid voltage with additional reactive current [1-2]. Figure 1(a) shows a typical voltage supportability curve that a WEGU have to be able to meet during low voltage ride through (LVRT) operation, while Figure 1(b) shows the principle of voltage back up by reactive current support. Among the motivations for the system operators to put in place such requirements are (i) to minimize the loss of generation during grid faults mitigating in this way the mismatches between demand and generation and (ii) to improve the network voltage profile.

The voltage dips due grid faults are classified in seven different types [4], Figure 2 shows the phasorial representation of each possible voltage unbalance due grid faults. The type A voltage dip is results from balanced three-phase short-circuits. All the other six types result from unbalanced faults that give rise to the appearance of voltage negative sequence components. The later are caused by the phase-to-ground (type B), phase-to-phase (type E) and the two-phase-to-ground (type C) faults. On the other hand, types F, G and D arise due the propagation of unbalanced faults through Δ-Y transformers, as shown Figure 2 [5]. Note that the type G occurs only when a two-phase-to-ground faults propagates through two series connected Δ-Y transformers.

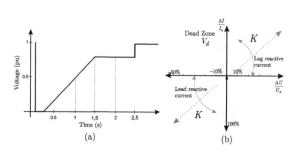

Figure 1. (a) LVRT curve, (b) Static characteristic for back up voltage operation [3].

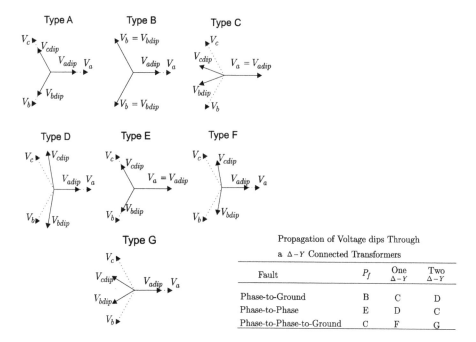

Figure 2. Voltage Dips Classification.

A significant voltage deviation is defined as a voltage deviation with a magnitude greater than the voltage dead band V_d as shown in Figure 1(b). Nowadays, the great majority of the grid codes states that WEGU must stay connected and execute the voltage back up during three-phase balanced significant voltage deviations. However, grid faults give rise to unbalanced voltage sags in the most cases. So far, voltage support is not always a requirement. However, in recent grid codes [3], WEGU have to be capable to use at least 40% of their nominal current capacity to reactive current support during unbalanced voltage dips. The negative sequence voltage brings some challenges for the WEGU operation, mainly concerning with the grid synchronization and the current control loop [6]. The grid synchronization has to be sufficiently fast and accurate to extract the negative and positive voltage sequences precisely, in order to generate the output current references. In [7] a synchronization method based on Kalman filter is presented for grid connected converters. This grid synchorization method has a good performance during balanced and unbalanced voltage dips. Another options are the traditional synchronous reference frame phase lock loop (PLL). In [8] are presented some variations on the traditional PLL for grid synchronization, making possible to extract the grid voltage sequence components even operating under unbalanced voltage conditions. On the other hand, the current controller has to be able to track negative sequence components. The conventional Synchronous Reference Frame Current Controller (SRFCC) ussualy presents a poor performance in these situations. The introduction of a double synchronous reference frame, synchronized with the positive and negative sequences, make possible to decouple the disturbance of the negative sequence on positive sequence current control loop and vice versa [9]. In [10] a current reference computation scheme was proposed in order to suppress the active power oscillations during unbalanced voltage sags as well as to avoid variations on the dc-link voltage. A review of current controller structures, in stationary and synchronous reference frames, is presented in [8]. Finally, if the current controller and the grid synchonization have a satisfactory performance, the impact of the WEGU on the grid directly dependents on how the current reference are defined. Many current reference methods for grid connected converters have been proposed aiming to reduce active and reactive power oscillations [11-12]. Nevertheless, emerging grid codes require voltage back up by reactive current injection during balanced and unbalanced voltage dips, where other issues, as the converter output current limitation for example, become more relevant than power oscillations supression. In addition, the injection of unbalanced currents during the LVRT operation may bring some useful effects as maximize the utilization of the converter current capacity and the reduction of the grid voltage unbalanced factor. However, depending on current reference methods, the effects may vary significantly [12].

In this chapter, a grid voltage negative sequence minimization (NSM) technique is proposed. It makes possible to inject unbalanced currents without exceeding the converter current ratings. To accomplish that, the proposed technique splits the current references in three parts. The first one is associated with the active power synchronized with the positive sequence voltage. The second one is related with the reactive power synchronized with the positive sequence while the third one is related with the reactive current synchronized with the negative sequence voltage. With the NSM technique, a simple and generic expression, depending only on the power references and voltages magnitudes information is derived to determine the possible amount of reactive power that can be added during an unbalanced voltage dip without overpass the converter current limits.

This chapter is organized as follow: Section II introduces some current reference methods used in the grid connected WEGU. Section III describes the proposed NSM method and the current limitation procedure. Section IV presents simulation results of the proposed method applied to a grid connected 2MW WEGU. Section V gives experimental results obtained in a grid connected converter of 10kW under unbalance voltage dips. Finally, section VI summarizes the main topics covered in this chapter.

2. System description

A grid connected WEGU with synchronous generator and full power converter is shown in Figure 3. During an unbalanced voltage dip, positive and negative sequence voltages components appear on the grid side converter terminals. Let us consider the grid voltages vector defined as:

$$\mathbf{v} = \begin{bmatrix} v_a & v_b & v_c \end{bmatrix},\tag{1}$$

where v_a, v_b and v_c are the WEGU grid side phase-to-ground voltages. On the same way, the converter output three-phase currents can be defined as:

$$\mathbf{i} = \begin{bmatrix} i_a & i_b & i_c \end{bmatrix}.\tag{2}$$

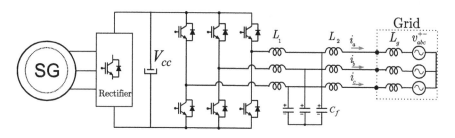

Figure 3. WEGU grid side converter.

In this chapter, the grid voltages and converter output currents are expressed in terms of their positive and negative sequence components at the fundamental frequency. In a three-phase three-wire system the voltages and currents vectors can be written as:

$$\mathbf{v} = \mathbf{v}^+ + \mathbf{v}^-\tag{3}$$

$$\mathbf{i} = \mathbf{i}^+ + \mathbf{i}^-\tag{4}$$

where the entries of $\mathbf{v}^+$ and $\mathbf{v}^-$ are the phase-to-ground sequence voltages. The $\mathbf{i}^+$ and $\mathbf{i}^-$ are composed by the output line positive and negative currents respectively. The instantaneous active and reactive power can be obtained by the dot product and vector product of **v** and **i**, which results:

$$p = \mathbf{v}^{+}\mathbf{i}^{+} + \mathbf{v}^{+}\mathbf{i}^{-} + \mathbf{v}^{-}\mathbf{i}^{+} + \mathbf{v}^{-}\mathbf{i}^{-} \tag{5}$$

$$q = \mathbf{v}^{+}_{\perp}\mathbf{i}^{+} + \mathbf{v}^{+}_{\perp}\mathbf{i}^{-} + \mathbf{v}^{-}_{\perp}\mathbf{i}^{+} + \mathbf{v}^{-}_{\perp}\mathbf{i}^{-} \tag{6}$$

where $\mathbf{v}_{\perp}$ is the voltage vector composed by the 90° lagged voltages in relation to the phase-to-ground voltages of vector **v**.

A classical solution of the reactive current support during unbalanced voltage dips is the injection of only positive sequence currents, following the same strategy used in balanced voltage dips [14]. This procedure may produce undesirable overvoltages on the non-affected phase and increase the unbalance factor between the grid voltages. On the other hand, during unbalanced voltage dips, it is possible to use the additional converter current capacity to mitigate the voltage unbalance. During the LVRT operation, the unbalanced output currents have to be controlled accurately, in order to avoid the converter to trip. On the next section, some basic selected current reference generation strategies are summarized to introduce the proposed negative sequence minimization reference current technique.

2.1. Literature review

A review of the main strategies that synthesizes with simplicity the current reference methods for grid connected converters during unbalanced voltage dips [8] [12] are presented in the following.

2.1.1. Instantaneous Active Reactive Control (IARC)

The most intuitive way to compute the current references during voltage dips is using the vector of voltages **v** and $\mathbf{v}_{\perp}$ and the active and reactive power references P^{*} and Q^{*}. The current reference vector can be obtained as:

$$\mathbf{i}^{*}_{p} = \frac{P^{*}}{|\mathbf{v}|^{2}}\mathbf{v} \tag{7}$$

$$\mathbf{i}^{*}_{q} = \frac{Q^{*}}{|\mathbf{v}|^{2}}\mathbf{v}_{\perp} \tag{8}$$

$$\mathbf{i}^{*} = \mathbf{i}^{*}_{p} + \mathbf{i}^{*}_{q} \tag{9}$$

The reference current vector $\mathbf{i}^*$ is composed by the sum of the active power current vector $\mathbf{i}_p^*$, where the components are in phase with the voltages on $\mathbf{v}$, and the reactive power current $\mathbf{i}_q^*$, where the components are in phase with $\mathbf{v}_\perp$.

In the case of balanced phase voltages, $\mathbf{v}=0$ and as a result the norm of the vector $|\mathbf{v}|$ is a constant value leading to sinusoidal balanced output currents. However, if some unbalance is present, $|\mathbf{v}|$ has an oscillation on twice de fundamental frequency. As a consequence, the current references will have and undesired harmonic components. Figure 4 (a) shows the resulting current references during a type B voltage dip at the terminals of a 1MW WEGU. Note that the active power is reduced during the fault, to allow the reactive current support. Before and after the voltage dip, only active power is delivered to the grid. It is possible to note that the active and reactive power are kept constant at the price of introducing harmonics on the output currents.

2.1.2. Balanced Positive Sequence (BPS)

In order to get only sinusoidal and balanced currents on the converter output, the current references are synchronized with the positive sequence voltage vector $\mathbf{v}^+$.

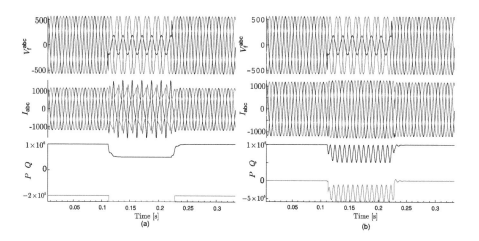

Figure 4. Reference currents and output power obtained during a type B unbalanced voltage dip to a a line voltage of 690Vrms, (a) references currents obtained using IARC and resulting output powers, (b) references currents obtained using BPS and resulting output powers.

The $\mathbf{i}_p^*$ and $\mathbf{i}_q^*$ reference currents are given by:

$$i_p^* = \frac{P^*}{\left|\mathbf{v}^+\right|^2}\mathbf{v}^+ \tag{10}$$

$$i_q^* = \frac{Q^*}{\left|\mathbf{v}^+\right|^2}\mathbf{v}_\perp^+ \tag{11}$$

Figure 4(b) shows the BPS reference currents obtained during type B voltage dip. The currents are always balanced even during the unbalanced voltage dips. The output powers are oscillatory due the interaction of the negative sequence voltage and the positive sequence currents.

2.1.3. Positive and Negative Sequence Control (PNSC)

This strategy allows the control of the positive and negative sequence currents associated with the active and reactive powers. The i_p^* and i_q^* vectors are obtained as follows:

$$i_p^* = \frac{P^*}{\left|\mathbf{v}^+\right|^2 - \left|\mathbf{v}^-\right|^2}(\mathbf{v}^+ - \mathbf{v}^-) \tag{12}$$

$$i_q^* = \frac{Q^*}{\left|\mathbf{v}^+\right|^2 - \left|\mathbf{v}^-\right|^2}(\mathbf{v}_\perp^+ - \mathbf{v}_\perp^-) \tag{13}$$

Figure 5 shows the PNSC resulting currents during a type B voltage dip. This strategy allows combine a set of positive and negative sequence voltages with the objective of cancelling the active power oscillations.

As seen, there are many possibilities to define the reference current. The injection of distorted currents in the IARC is undesired since generation units usually have to avoid the injection of harmonic content into the grid, even during the LVRT operation due the power quality issue. The BPS method overcomes current distortion issue since it injects only balanced currents to the voltage back-up. As a consequence, it allows boosting the grid positive sequence voltage if required. On the other side, the PNSC strategy shows the possibility to combine the negative sequence components to eliminate output power oscillations, resulting in unbalanced output currents. However, the important point of the mitigation of negative sequence voltage is not addressed. The next section presents the NSM technique, where the positive and negative sequence voltages are used to inject reactive current during the LVRT operation.

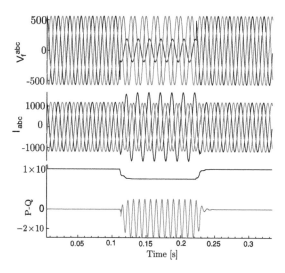

Figure 5. Reference currents and output active and reactive powers during a type B voltage dip. Gains k_p=1 and k_q=1 for balanced currents and k_p=1 and k_q=0.5 for unbalanced currents.

3. Negative Sequence Voltages Minimization (NSM) technic

During the LVRT operation, the reactive power reference Q^* is obtained as shown in equation (15), where I_q follows the reative current support rules stated in [3]

$$I_q = \frac{K(V^+ - \bar{V}_1^+) - V_d}{V_n} I_n.$$ (14)

Gain K is adjustable from 1 to 10, depending on an agriment with the grid utility operator, V^+ is the positive sequence voltage immediatly after the voltage dip occurs, $\bar{V}_1^+$ is the 1-minute average value of the positive sequence voltage prior to the voltage dip, V_n is rated line-to-line Root-Mean-Square RMS voltage, V_d is the voltage deadband and I_n is the rated RMS line current of the WEGU.

As a result, the reactive power reference Q^* during the LVRT operation can be expressed as:

$$Q^* = 3V^+ I_q$$ (15)

According to the German grid code [3], in three-phase voltage dips, the WEGU has to be able to use up to 100% of its current rating capacity to the voltage support, accordingly with Figure 1(b). During a significant voltage deviation, the current associated with the active power, has to be reduced in the benefit of the reactive current feed-in. In the case of unbalanced voltage dips, the significant voltage deviation is detected if the $V^+ < 0.9 V_n$, in this case the WEGU must be technically able to feed in a reactive current of at least 40% of the rated current. In [14], a WEGU operation mode during significant voltage deviations is demonstrate, the current injected are balanced and in phase with the positive sequence voltages only.

On the unbalanced voltage dips case, 40% of the current capacity has to be used during the reactive current support. There is no obligation to use the remaining current capacity. In addition, it is well know that the utilization factor of a typically WEGU range from 20% to 40% [15]. A good alternative in these cases is injecting reactive current synchronized with the negative sequence in order to mitigate unbalance between the voltages.

As a consequence, the active and reactive current references, to normal operation and during voltage dips can be defined as

$$i_p^* = \frac{P^*}{\left|v^+\right|^2} v^+ \tag{16}$$

$$i_q^* = \frac{Q^*}{\left|v^+\right|^2} v_\perp^+ + \frac{B^*}{\left|v^-\right|^2} v_\perp^- \tag{17}$$

where B^* is the negative sequence reactive power reference. Note that in equation (16), the negative sequence voltage is not included. The problem that arises is how to find B^* in order to guarantee that the output current do not overpass a pre-defined limit. In the next subsection, a current limitation algorithm is derived.

3.1. Current limitation

Let us assume that the output limit current of a grid-connected converter is the same in each phase. The current amplitude range in each phase can be defined in the interval $i_x = \left[-I_x^{lim}, \ I_x^{lim}\right]$, where $x = a, b, c$. In addition, as a three wire converter is considered, the sum of all the three output currents is always zero. Considering this conditon, is possible to define a set of possible output currents and gathering them into a vector i_{abc}.

To obtaine the currents in the $\alpha\beta$ reference frame, the well-known abc-$\alpha\beta$ transformation is applied to the all the possible i_{abc} vector.

$$
\mathbf{i}_{\alpha\beta} = \sqrt{\frac{2}{3}}
\begin{bmatrix}
1 & -\dfrac{1}{2} & -\dfrac{1}{2} \\[2mm]
0 & \dfrac{\sqrt{3}}{2} & -\dfrac{\sqrt{3}}{2} \\[2mm]
\dfrac{1}{\sqrt{2}} & \dfrac{1}{\sqrt{2}} & \dfrac{1}{\sqrt{2}}
\end{bmatrix}
\mathbf{i}_{abc}. \tag{18}
$$

where $\mathbf{i}_{\alpha\beta}$ is the current vector in the $\alpha\beta$ reference frame.

The mapping of all possible $\mathbf{i}_{abc}$ vectors in the $\alpha\beta$ reference frame are confined within a hexagon, as show in Figure 6. Hence, to guarantee that the output currents are limiting in the abc-framework, is necessary to limit the $\alpha\beta$ reference currents inside the hexagon. In a normal operation situation, i_a and i_β are balanced and sinusoidal, consequently the $\mathbf{i}_{\alpha\beta}$ vector describes a circle inside the hexagon.

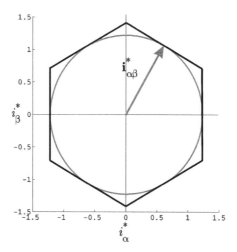

Figure 6. (a) Converter current capability curve in $\alpha\beta$ coordinates.

To develop a methodology to compute B^*, and assures that the output currents stay inside the hexagon, it is necessary to analyze the behavior of the output currents obtained by equations (16) and (17). Assuming that the phase voltages on vector $\mathbf{v}$ in (1) can be expressed as:

$$
\begin{aligned}
v_a &= v_a^+ + v_a^- \\
v_b &= v_b^+ + v_b^- \\
v_c &= v_c^+ + v_c^-
\end{aligned}
\tag{19}
$$

where δ_v is the voltage angle and V_{abc} are the phase magnitudes.

The synchronization system presented in [7] filters the fundamental frequencies of the phase voltages of (19). In addition, the 90° lagged components of each phase voltage are also available. Applying the transformation in [16], the positive and negative sequence voltages can be expressed as:

$$v^+_a = V^+ \sin(\omega t + \delta^+_v) \tag{20}$$

$$v^+_b = V^+ \sin(\omega t + \delta^+_v - 120°) \tag{21}$$

$$v^+_b = V^+ \sin(\omega t + \delta^+_v + 120°) \tag{22}$$

$$v^-_a = V^- \sin(\omega t + \delta^-_v) \tag{23}$$

$$v^-_b = V^- \sin(\omega t + \delta^-_v + 120°) \tag{24}$$

$$v^-_b = V^- \sin(\omega t + \delta^-_v - 120°) \tag{25}$$

The 90° lagged components of each phase voltage are also available:

$$v^+_{aq} = V^+ \sin(\omega t + \delta^+_v + 90°) \tag{26}$$

$$v^+_{bq} = V^+ \sin(\omega t + \delta^+_v - 30°) \tag{27}$$

$$v^+_{cq} = V^+ \sin(\omega t + \delta^+_v + 210°) \tag{28}$$

$$v^-_{aq} = V^- \sin(\omega t + \delta^-_v - 90°) \tag{29}$$

$$v^-_{bq} = V^- \sin(\omega t + \delta^-_v + 30°) \tag{30}$$

$$v_{cq}^- = V^- \sin(\omega t + \delta_v^- - 210^\circ) \tag{31}$$

where δ_v^+ and δ_v^- are the positive and negative reference angles in relation to δ_v. Using the abc-$\alpha\beta$ transformation to get the voltages of (19) in the $\alpha\beta$f ramework, we obtain

$$\begin{bmatrix} v_\alpha \\ v_\beta \end{bmatrix} = T_{\alpha\beta} \begin{bmatrix} v_a \\ v_b \\ v_c \end{bmatrix}. \tag{32}$$

The same procedure can be applied to (20-31), resulting in the following variables respectively v_α^+, v_β^+, v_α^-, v_β^-, $v_{\alpha\perp}^+$, $v_{\beta\perp}^+$, $v_{\alpha\perp}^-$ and $v_{\beta\perp}^-$.

The reference currents in the $\alpha\beta$ framework, accordingly with the equations (16) and (17) can be expressed as

$$i_\alpha^* = i_{\alpha p}^+ + i_{\alpha q}^+ + i_{\alpha q}^- \tag{33}$$

$$i_\beta^* = i_{\beta p}^+ + i_{\beta q}^+ + i_{\beta q}^- \tag{34}$$

where the positive sequence currents are determined by

$$i_{\alpha p}^+ = \frac{P^*}{\left| \mathbf{v}^+ \right|} v_\alpha^+ \tag{35}$$

$$i_{\beta p}^+ = \frac{P^*}{\left| \mathbf{v}^+ \right|} v_\beta^+ \tag{36}$$

$$i_{\alpha q}^+ = \frac{Q^*}{\left| \mathbf{v}^+ \right|} v_{\alpha\perp}^+ \tag{37}$$

$$i_{\beta q}^+ = \frac{Q^*}{\left| \mathbf{v}^+ \right|} v_{\beta\perp}^+. \tag{38}$$

The negative sequence voltages are given by:

$$i_{\alpha q}^- = \frac{B^*}{\left|\mathbf{v}^-\right|} v_{\alpha\perp}^- \tag{39}$$

$$i_{\beta q}^- = \frac{B^*}{\left|\mathbf{v}^-\right|} v_{\beta\perp}^-. \tag{40}$$

In the balanced case, the reference currents are in phase with the positive sequence voltage. In the case of $P^*>0$, $Q^*\neq0$ and $B^*=0$, the representation of the current vector $\mathbf{i}_{\alpha\beta}^*$ is a circle. On the other hand, if B^* is not null, reactive power currents synchronized with the negative sequence voltage are added in the reference currents of equation (33-34). As a consequence, the graphically representation of the reference currents describes an ellipse on the $\alpha\beta$ plane. Figure 7 shows the current limit circle and an inscribed ellipse. It is important to observe that i_α and i_β currents do not reach its maximum possible values, but the $|\mathbf{i}_{\alpha\beta}^*|$ reaches the circle in some point, producing the maximum allowed output current on the abc-framework.

Taking equations (35-40), is possible to define the constants k_2, k_3 and k_1:

$$k_1 = \frac{P^*}{\left|\mathbf{v}^+\right|} \tag{41}$$

$$k_2 = \frac{Q^*}{\left|\mathbf{v}^+\right|} \tag{42}$$

$$k_3 = \frac{B^*}{\left|\mathbf{v}^-\right|} \tag{43}$$

Replacing equations (41-43) in (33-34) and writing the $\alpha\beta$ references components in sine and cosine, the current references can be expressed as:

$$i_\alpha^* = k_1 \sin(\omega t + \delta_v^+) - k_2 \cos(\omega t + \delta_v^+) - k_3 \sin(\omega t + \delta_v^-) \tag{44}$$

$$i_\beta^* = k_1 \cos(\omega t + \delta_v^+) + k_2 \sin(\omega t + \delta_v^+) + k_3 \cos(\omega t + \delta_v^-) \tag{45}$$

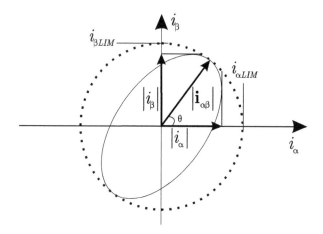

Figure 7. Negative sequence component effect on the norm of currents vector $|\mathbf{i}_{\alpha\beta}|$.

The challenge is to find an expression that defines the norm of a given set of current references i_α^* and i_β^*. Making $\omega t + \delta_v^+ = \theta$ and $\omega t + \delta_v^- = \theta + \theta_-$ on equations (44) and (45) results

$$i_\alpha^* = k_1 \cos(\theta) + k_2 \sin(\theta) + k_3 \cos(\theta + \theta_-) \tag{46}$$

$$i_\beta^* = k_1 \sin(\theta) - k_2 \cos(\theta) - k_3 \sin(\theta + \theta_-) \tag{47}$$

In order to find the norm of vector $\mathbf{i}_{\alpha\beta}^*$, considering the current references expressed in (46-47), is possible to write that

$$\left\| \mathbf{i}_{\alpha\beta}^* \right\| = \sqrt{(k_1 \sin(\theta) - k_2 \cos(\theta) - k_3 \sin(\theta + \theta_-))^2 + (k_1 \cos(\theta) + k_2 \sin(\theta) + k_3 \cos(\theta + \theta_-))^2}. \tag{48}$$

Taking the derivative in function of θ and equal to zero, is possible to find a point of maximum of the ellipse:

$$\frac{d}{d\theta}\sqrt{(k_1\sin(\theta)-k_2\cos(\theta)-k_3\sin(\theta+\theta_-))^2+(k_1\cos(\theta)+k_2\sin(\theta)+k_3\cos(\theta+\theta_-))^2}=0 \qquad (49)$$

Simplifying the result of equation (49), results in:

$$4k_2k_3\cos(2\theta+\theta_-)-4k_1k_3\sin(2\theta+\theta_-)=0. \qquad (50)$$

Solving equation (50) in terms to θ:

$$\theta=\frac{\tan^{-1}\left(\dfrac{k_2}{k_1}\right)-\theta_-}{2} \qquad (51)$$

Substituting (51) in (48) results an expression that determines the norm of $\mathbf{i}^*_{\alpha\beta}$ in function of the k_1, k_2 and k_3 parameters:

$$\left\|\mathbf{i}^*_{\alpha\beta}\right\|=\sqrt{k_1^2+k_2^2+k_3^2+2k_3\sqrt{k_1^2+k_2^2}}. \qquad (52)$$

An important result of (52) is the possibility to get the maximum norm without taking use of the angles δ_v^+ and δ_v^-. Making $\|\mathbf{i}^*_{\alpha\beta}\|=I_{\alpha\beta}^{\lim}$ is possible to solve equation (52) to k_3. The resulting equation gives us two possible solutions

$$k_3^{+-}=\frac{-2k_1^2-2k_2^2\pm2I_{\alpha\beta}^{\lim}\sqrt{k_1^2+k_2^2}}{2\sqrt{k_1^2+k_2^2}}. \qquad (53)$$

Only the positive solution for equation (53) is valid, once that $k_3\geq0$ for the reactive capacitive power injection. So, B^* can is given by

$$B^*=k_3^+\left|\mathbf{v}^-\right|. \qquad (54)$$

Figure 8(a) shows the $\|\mathbf{i}^*_{\alpha\beta}\|$ trajectory in the $\alpha\beta$ plane for $k_3=0$. Otherwise, in Figure 8 (b), (c) and (d), presents the resulting ellipse and the circle with the norm obtained by equation (52). The maximum norm is finding accurately, independent on the voltage phase angles. The angle θ depends on the active and reactive power references and on the negative sequence angle δ_v^-, as shown in equation (51).

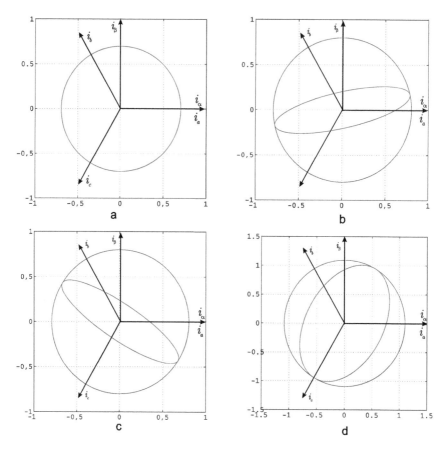

Figure 8. (a) $\alpha\beta$ framework current norm for $k_1=0.7I_n$, $k_2=0$ e $k_3=0$, (b) Circle with radius computed by (42) and re-sulting ellipse for $k_1=0$, $k_2=-0.5I_n$, $k_3=0.3I_n$ $\theta=40°$, (c) Circle and ellipse for $k_1=0$, $k_2=0.5I_n$, $k_3=0.3I_n$ e $\theta=40°$, (d) Cir-cle and ellipse for $k_1=0.6I_n$, $k_2=0.5I_n$, $k_3=0.3I_n$ e $\theta=40°$.

4. Simulation results

In order to demonstrate the NSM current limitation algorithm, a WEGU system composed by a three phase, three wire converter, is connected to the electrial network show in Figure 9. The WEGU is connected on the low voltage side of the Transformer T3 on a line to line voltage of 690 Vrms and with a nominal power of $P_n=2MW$. The short circuit ratio on the PCC is near to 4, what represent a weak grid model. The voltage source converter male use of a synchro-nization system base on Kalman Filter, as presented in [7]. The current controllers are imple-mented in stationary frames and are based in resonant controllers as presented in [17]. All the simulations are executed in Matlab/Simulink ®. To demonstrate the NSM algorithm perform-ance a two-phase-to-ground fault on the point (P1) in the distribution feeder will be tested. Figure 10 shows the resulting variables. Prior the fault, $t < T_1$, the WEGU is injecting $0.2P_n$ into the grid. At T_1 the fault occurs, and the WEGU LVRT operation initiate. In order to meet the grid code requirements [3], 40% of the current converter capacity is used to inject balanced reactive currents using the BPS strategy.

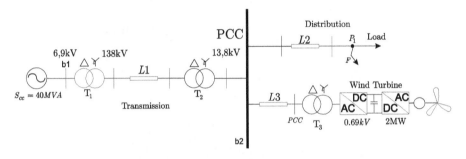

Figure 9. Electrical network for the evaluation of the NSM model.

On Fig. 10(c) is possible to see how the current support boost the positive sequence voltage from 388V to 497V at the PCC. However, the negative sequence almost remaing the same, changing from 161V to 165V. At time T_2, the current reference strategy is switched to the NSM. Equation (53) computes the value of k_3 in order to the pre-defined $I_{\alpha\beta}^{lim}$. Is possible to see that output current i_a reach the maximum allowed current, in this case setted to 2366 A. The positive sequence voltage remains the same as in $t < T_2$, but the negative sequence voltage reduces from 165V to 67V. The unbalanced factor on the PCC voltages of Figure 10(b) is clearly lower. One of the more relevant result of this work can be infered on Figure 10(d). The magnitude of the current reference vector $\mathbf{i}^*_{\alpha\beta}$ is computed by equation (52) during all the simulation. It can be seen that the result is valid to balanced and unbalanced currents. Between times T_2 and T_3, the currents are unbalanced and the value of $\|\mathbf{i}^*_{\alpha\beta}\|$ does not reach the current references peak value as in the balanced case. Figure 11 depictes the graphically trajectory of $\mathbf{i}^*_{\alpha\beta}$ during the balanced (BPS) and unbalanced (NSM) operation. The ellipse trajectory does not overpass the current limit circle what guarantees the current limitation on the abc framework. In addition, is worth to point out that i^*_α and i^*_β will reach its maximum values only if the resulting ellipse is aligned with the α or β axis. At T_3 the fault is cleared and the reactive current support is finished, the active power reference stays in $0.2P_n$ during all the simulation time.

An important parameter to measure the level of unbalance voltages is the unbalanced factor [13] given by:

$$\%VUF = \frac{\left|v^-\right|}{\left|v^+\right|} \times 100 \tag{55}$$

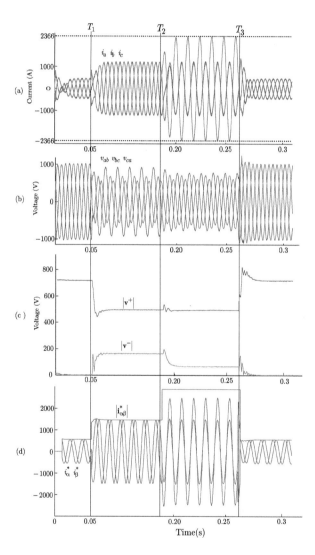

Figure 10. BPS and NSM strategies during an unbalanced voltage dip on weak grid conditions: (a) Grid connected converter output currents, (b) PCC voltages. (c) Positive and negative sequence voltages magnitude, (d) Reference currents and $\mathbf{i}^*_{\alpha\beta}$ computed by equation (52).

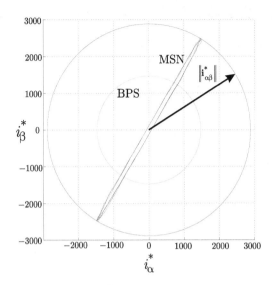

Figure 11. Reference current trajectory in the current spaces.

In order to compare the performance of the current reference strategies, a set of simulations to evaluate the unbalance factor are investigated. Two diferrent faults on the distribuiton feeder (P1) are considered. Fault F1 is a phase-to-ground fault that results in a type C voltage dip on the converter terminals. Fault F2 is a phase-to-phase to ground fault, as a consequence a type F voltage dip arise on the WEGU terminals. Another two issues are important to be considered. The active power level on the moment of the fault and the short circuit ratio on the PCC. To F1 and F2, two levels of active power are considered: $0.2P_n$ and $0.5P_n$. Above this active power level, and taking into account that 40% of the reactive current support is done by balanced currents, there will be no much remaing current capacity available to inject negative sequence reactive current. In addition, the transmisison line L1 and distribution line L3 have their lenghts reduced in order do modify the short circuit ratio on the PCC to 10. In all this simulations, the back up voltage support rulles of [3] are followed. Table 1 shows the resulting unbalance factor for the simulations using BPS, NSM or PNSC current reference strategies. By the analysis of the simulations results, some conclusions can be inferred:

- The BPS has a poor performance in all situation, when compared with the PNSC and NSM strategies;

- The type of faults does not change the effectiveness of the strategies;

- On the $0.5P_n$ active power level, the PNSC and NSM have similar performances in the strong and weak grid conditions;

- The NSM has a better performance in the $0.2P_n$ active power level, specialy in the weak grid condition.

In the next section, some experimental results of the NSM strategy are presented

Type of Fault	Phase-to-Ground (F1)				Phase-to-Phase to Ground (F2)			
Short Circuit Ratio	4		10		4		10	
Strategies\Active Power	0.2Pn	0.5Pn	0.2Pn	0.5Pn	0.2Pn	0.5Pn	0.2Pn	0.5Pn
PNSC	19,4%	18,3%	19,5%	19,4%	22,7%	20,2%	25,6%	29,8%
BPS	25%	23,5%	24,4%	23,1%	33,2%	30,6%	34,8%	33,8%
MSN	9,9%	18,6%	11,4%	16%	13,2%	23,3%	23,6%	29,6%

Table 1. Resulting unbalanced Factors.

5. Experimental results

The experimental setup is composed by a three-phase PWM voltage-source converter inveter of 10-kVA, and an impedance switching voltage sag generator (VSG) connected to the grid as shown in Fig. 12. The synchornization system extract the sequence components by filtering the grid phase voltages. Equation (15) gives the positive sequence reactive power reference Q^* and equation (40) gives the negative sequence power reference B^*. The switching frequency is $f_s = 10kHz$, inductors $L_1 + L_2 = 2mH$, $C_f = 40\mu F$, $V_{dc} = 200V$ and line to line voltage of 100 V_{rms}.

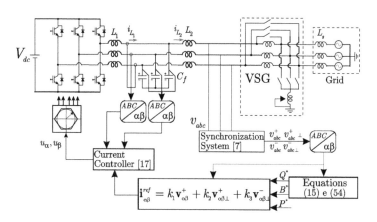

Figure 12. Experimental setup.

Considering the grid connected inverter injecting 500W into the grid prior the fault. The VSG is set to perform a phase-to-phase-to ground fault, with duration of 500ms, resulting in a type C voltage dip. The impedances are set to provide a reduction on the positive sequence voltage of 35%. Figure 13 (a) shows the resulting line-to-line voltages on the on the converter terminals with a peak value of 152V. The positive sequence voltage decrease to 121V during the fault and the negative sequence voltage arise to 27V. Figure 13(b) shows the converter output currents and the power references before, during and after the voltage dip. Before the fault, the converter is only injecting active power into the grid. As the unbalanced voltage dip initiate, the NSM strategy compute the reactive power references by equations (15) and (54). The current limit is set to 20 A, i_a current reachs this limit as shows the red line limit.

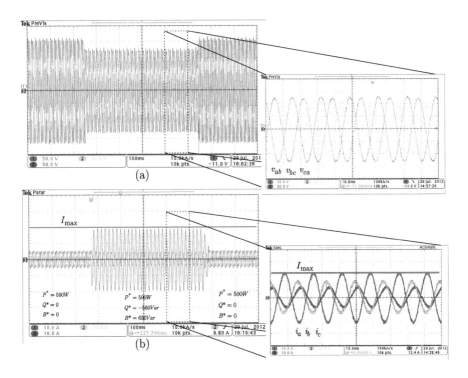

Figure 13. Experimental results, (a) line to line voltages on the PCC, (b) output currents.

6. Conclusions

This chapter has presented a new control method to limit the output current grid connected converter during unbalanced voltages sags voltage sags. WEGU are required to contribute for the voltage back up after balanced and unbalanced voltage dips. However, during unbalanced voltage dips negative sequence voltages arise on the wind turbine terminals. The voltage support by reactive current support is required, and the WEGU has to use at least 40% of their current capacity to back up the grid voltage. The injection of positive sequence reactive current during faults permits boosting the voltage at the PCC. In addition, the injection of negative sequence reactive power allows reduce the unbalance between the voltages. Hence, if unbalanced currents may be injected into the grid, to assure that the converter do not disconnect during the LVRT, is necessary to limit the output currents. This work proposes a current reference generation called negative sequence minimization (NSM). With this method is possible to find the maximum reactive power references, in order to keep the output current limited. For instance, given an active power reference and a set of positive and negative sequence voltages, related to a generic unbalanced voltage dip, is possible to find the reactive power references without exceeding the maximum output converter output current. Another advantage of this strategy is that the grid voltage angles are not required to implement the algorithm.

Simulation and experimental results have shown the feasibility and the effectiveness of the proposed method.

Author details

Ivan Jorge Gabe [1*], Humberto Pinheiro[2] and Hilton Abílio Gründling[2]

*Address all correspondence to: ivangabe@gmail.com

1 Federal Institute of Rio Grande do Sul-IFRS, Farroupilha, Brazil

2 Federal University of Santa Maria, Santa Maria, Brazil

References

[1] Tsili, M, & Papathanassiou, S. A review of grid code technical requirements for wind farms. IET Renewable Power Generation (2009). , 3(3), 308-332.

[2] Alegría, I. M, Andreu, J, Martín, J L, Ibañez, P, Villate, J L, & Camblong, H. Connection requirements for wind farms: A survey on technical requirements and regulation. In ELSEVIER Renewable and Sustainable Energy Reviews, (2007). , 11(8), 1858-1872.

[3] Ministry for the EnvironmentNature Conservation and Nuclear Safety. "Ordinance on System Services by Wind Energy Plants" (System Service Ordinance-SDLWindV)". http://www.erneuerbareenergien.de/files/pdfs/allgemein/application/pdf/sdl_windv_en.pdf.accessed 3 July (2012).

[4] Bollen, M, & Zhang, J. L D. Different methods for classification of three-phase unbalanced voltage dips due to faults. In ELSEVIER Renewable and Sustainable Energy Reviews, (2003). , 11(1), 59-69.

[5] Bollen, M, & Zhang, J. L D. Characteristic of Voltage Dips (Sags) in Power Systems. IEEE Transactions on Power Delivery, (2000). , 13(2), 827-832.

[6] Gabe, I J, & Pinheiro, H. Impact of Unbalance Voltage Dips on the Behaviour of Voltage Source Inverters. In proceedings of Brazilian Power Electronics Conference (2009). COBEP'09, Sept. 27 2009-Oct. Bonito-MS Brazil., 1, 2009-956.

[7] Gabe, I J, Palha, F K, & Pinheiro, H. Grid Connected Voltage Source Inverter Control During Voltage Dips. In proceedings of IEEE Annual Conference 35th Industrial Electronics, (2009). IECON'Nov 2009, 4571- 4576., 09, 3-5.

[8] Teodorescu, R, Liserre, M, & Rodrígues, P. Grid Converters for Photovoltaic and Wind Power Systems. Wiley April (2011).

[9] Song, H S, & Nam, K. Dual Current Control Scheme for PWM Converter under Unbalanced Input Voltage Conditions. IEEE Transactions on Industrial Electronics (1999). , 46(5), 953-959.

[10] Rioual, P, Pouliquen, H, & Louis, J. P. Regulation of a PWM Rectifier in the Unbalanced Network State Using a Generalized Model. IEEE Transactions on Power Electronics (1996). , 11(3), 495-502.

[11] Wang, F, Duarte, J L, Hendrix, M, & Design, M. and Analysis of Active Power Control Strategies for Distributed Generation Inverters under Unbalanced Grid Faults. IET Generation, Transmission and Distribution (2010). , 4(8), 905-916.

[12] Rodriguez, P, Timbus, A. V, Teodeorescu, R, Liserre, M, & Blaabjerg, F. Independent PQ Control for Distributed Power Generation Systems under Grid Faults. In proceedings of 32° Annual IEEE Industrial Electronics Conference, IECON'06, (2006). Paris France.

[13] Lee, C T, Hsu, C H, & Cheng, P T. A Low-Voltage Ride-Through Technique for Grid-Connected Converters of Distributed Energy Resources. IEEE Transactions on Industry Application (2011). , 47(4), 1821-1832.

[14] Fischer, M, & Schellschmidt, M. Fault Ride Through performance of Wind Energy Converters with FACTS capabilities in response to up-to-date German grid connection requirements. In proceedings of European Wind Energy Conference and Exhibition, EWEC2010, (2010). Warsaw Poland.

[15] Brazilian National System Operator (ONS)Acompanhamento Mensal da Geração de Energia das Usinas Eolielétricas com Programação e Despacho Centralizados pela ONS (2012). http://www.ons.org.br/download/resultados_operacao/boletim_mensal_geracao_eolica/Boletim_Eolica_fev-2012.pdf.acessed on 27 july 2012).

[16] Fortescue, C L. Method of symmetrical co-ordinates applied to the solution of polyphase networks. Transaction of the Electrical Engineers (1928). , 37-1027.

[17] Gabe, I J, Montagner, V F, & Pinheiro, H. Design and Implementation of a Robust Current Controller for VSI Connected to the Grid Through an LCL Filter. IEEE Transactions on Power Electronics (2009). , 24(6), 1444-1452.

Chapter 5

Wind Farms as Negative Loads and as Conventional Synchronous Generation – Modelling and Control

Roberto Daniel Fernández,

Pedro Eugenio Battaiotto and Ricardo Julián Mantz

Additional information is available at the end of the chapter

1. Introduction

The concept of negative load [1,2] has been applied to wind generators to indicate their capability for delivering current meanwhile their voltage is imposed by the electrical system at the connection point. More recently, the same concept was applied for studying dispatch or spinning reserve considering that the total regulating power required at any moment depends on the sum the system load and the wind power which can counterbalance or increase load variations. In this way, aggregated variations must be investigated regarding wind power as negative load [3,4].

Traditionally, induction generators, squirrel cage and double fed (wound rotor) induction generators (DFIG) have been considered as current sources (or power ones) in power system analysis. Indeed, for analyzing power system stability in a linear frame, i.e. by small signal analysis, it is possible to find the power system eigenvalues and concluding about stability following the next steps:

1. Writing differential algebraic non linear model of the power system considering wind turbines wind induction machines as negative loads.

2. Linearizing the non linear model [2,5,6].

Checking the movement of the system eigenvalues when wind fixed speed generation is increased [7] or when different control strategies for active and reactive powers are applied to DFIGs wind farm [8,9].

Even when modelling wind generators or wind farms as current sources have shown through linear and non linear analysis [10] that wind farms can contribute to the power systems stability, it is important to consider that power systems have been developed from voltage synchronous generation, i.e. they (power systems) have not been developed with variable current sources as wind farms. For this reason, it is usually assumed that, as cited in [11], *'These plants (wind farm ones) exhibit static and dynamic characteristics that differ fundamentally from that of conventional generators. As a result, wind power plants do not fit the template for models of conventional generating facilities.'*

However, considering wind farms as (variable) voltage sources could help not just for a better understanding of induction machines in an electrical grid but also for mimicking conventional voltage source behaviors with wind power plants.

In this way, this chapter developes the equivalent model of induction generators representing them as a voltage sources with a series impedance. As a consequence, aside from the variability of wind, it would allow to analyze wind energy generation as another voltage source in power systems and, then, it should be possible to introduce "standard" rules for conventional generation to non conventional ones.

The structure of this chapter is as follows. Firstly, the concept of "general reference frame" is introduced in order to analyze squirrel cage and double fed induction machines. The dynamics model of the induction machine is deduced by considering a common simplification about (fast) stator dynamics. Then, equivalent Thevenin models are presented according to internal voltages sources and as a which are functions of the active and reactive powers. Secondly, it is shown that it is possible controlling active and reactive wind generator powers. Finally, a Thevenin agregated model of the wind farm is proposed and some stability concerns of power systems are considered. In this way, Lyapunov theory is applied looking for demonstrating wind farms contribution to the power system stability considering wind farms as currents sources but also as voltage ones. In this last regard some wind farm control rules are derived from exploiting similarities between conventional generation and wind farms.

2. Induction machine equations in different reference frames

This section relies in [12] for presenting induction machine dynamics from a two axes general reference frame.

It is well known that electrical machines solve the problem of obtaining a rotating field by employing three windings sinusoidally distributed and separated by 120º (mechanical degrees) which are fed by three sinusoidal stator 120° electrical degree phase shift. However, because of field distributions are the same along the third dimension (the machine shaft direction), these field distributions are analyzed in the plane where only are needed two linearly independent directions for characterizing any movement. The relationship between the three-phase (A, B and C for stator and a, b and c for rotor) and two phase voltages taking into account natural frames (fixed to the stator sD - sQ and fixed to the rotor rα - rβ for stator and rotor quantities, respectively) are:

$$\begin{bmatrix} u_{s0} \\ u_{sD} \\ u_{sQ} \end{bmatrix} = \frac{2}{3} \begin{bmatrix} 1/2 & 1/2 & 1/2 \\ 1 & -1/2 & -1/2 \\ 0 & \sqrt{3}/2 & -\sqrt{3}/2 \end{bmatrix} \begin{bmatrix} u_{sA} \\ u_{sB} \\ u_{sC} \end{bmatrix} \tag{1}$$

$$\begin{bmatrix} u_{r0} \\ u_{r\alpha} \\ u_{r\beta} \end{bmatrix} = \frac{2}{3} \begin{bmatrix} 1/2 & 1/2 & 1/2 \\ 1 & -1/2 & -1/2 \\ 0 & \sqrt{3}/2 & -\sqrt{3}/2 \end{bmatrix} \begin{bmatrix} u_{ra} \\ u_{rb} \\ u_{rc} \end{bmatrix} \tag{2}$$

By considering the two axes description, quantitative and qualitative analyses of induction machines can be simplified and also vector control concepts can be used. In this way, the contribution of vector control is based in controlling the induction machines active and reactive powers independently and/or controlling them as DC equivalent ones.

In a general reference frame, all induction machine variables are referred to a real axis known as direct axis x and to the quadrature axis y both rotating at the reference frame speed $\omega_g = d\theta_g / dt$ as shown in Figure 1. In this figure, θ_g is the angle of the real axis x meas‐ure from sD.

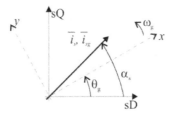

Figure 1. Stator current vector in a general reference frame

Then, the current stator phasor defined in the general framework, is:

$$\bar{i}_{sg} = \bar{i}_s e^{-j\theta_g} = i_{sx} + ji_{sy}$$

where upper bar indicates phasor quantities.

Also,

$$\bar{u}_{sg} = \bar{u}_s e^{-j\theta_g} = u_{sx} + ju_{sy}$$

$$\bar{\psi}_{sg} = \bar{\psi}_s e^{-j\theta_g} = \psi_{sx} + j\psi_{sy}$$

With $\bar{u}_s$ and $\bar{\psi}_s$ stator voltage and (linked) flux space phasors in the general reference frame.

At the rotor side, Figure 2 shows three frames, rotor ($r\alpha$ and $r\beta$), stator (sD and sQ) and general (x and y) and their angles θ_r, 0 and θ_g, respectively.

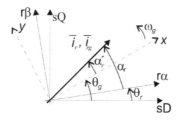

Figure 2. Different reference frames for the current rotor phasor

Meanwhile, the current phasor in the rotor reference frame can be expressed as $i_r = |i_r| e^{j\alpha_r}$, in the general reference frame is $i_{rg} = |i_r| e^{j\alpha_r'}$ with $\alpha_r' = \alpha_r - (\theta_g - \theta_r)$. Then,

$$i_{rg} = |i_r| e^{j\alpha_r} e^{-j(\theta_g - \theta_r)} = i_r e^{-j(\theta_g - \theta_r)} = i_{rx} + ji_{ry}.$$

Also, rotor voltage and (linked) flux space phasors in the same frame are:

$$\bar{u}_{rg} = \bar{u}_r e^{-j(\theta_g - \theta_r)} = u_{rx} + ju_{ry}$$

$$\bar{\psi}_{rg} = \bar{\psi}_r e^{-j(\theta_g - \theta_r)} = \psi_{rx} + j\psi_{ry}.$$

Finally, induction machine phasor expressions are:

$$\bar{u}_{rg} = R_r \bar{i}_{rg} + \frac{d\bar{\psi}_{rg}}{dt} + j(\omega_g - \omega_r)\bar{\psi}_{rg},$$

$$\bar{u}_{sg} = R_s \bar{i}_{sg} + \frac{d\bar{\psi}_{sg}}{dt} + j\omega_g \bar{\psi}_{sg}$$

being R_s and R_r stator and rotor resistances, respectively, and $\bar{u}_{rg} = 0$ when a squirrel cage machine is considered. Additionally, stator and rotor fluxes can be expressed in terms of current phasors, and the stator, rotor and magnetizing inductances (L_s, L_r, L_m respectively)

$$\bar{\psi}_{sg} = L_s \bar{i}_{sg} + L_m \bar{i}_{rg}$$

$$\bar{\psi}_{rg} = L_r \bar{i}_{rg} + L_m \bar{i}_{sg}.$$

Last four equations can be rewritten compactly in matrix form as follows:

$$\begin{bmatrix} \bar{u}_{sg} \\ \bar{u}_{rg} \end{bmatrix} = \begin{bmatrix} R_s & 0 \\ 0 & R_r \end{bmatrix} \begin{bmatrix} \bar{i}_{sg} \\ \bar{i}_{rg} \end{bmatrix} + \frac{d}{dt} \begin{bmatrix} L_s & L_m \\ L_m & L_s \end{bmatrix} \begin{bmatrix} \bar{i}_{sg} \\ \bar{i}_{rg} \end{bmatrix} + j\omega_g \begin{bmatrix} L_s & L_m \\ L_m & L_r \end{bmatrix} \begin{bmatrix} \bar{i}_{sg} \\ \bar{i}_{rg} \end{bmatrix} - j\omega_r \begin{bmatrix} 0 & 0 \\ L_m & L_r \end{bmatrix} \begin{bmatrix} \bar{i}_{sg} \\ \bar{i}_{rg} \end{bmatrix}, \qquad (3)$$

or by considering real and imaginary components:

$$
\begin{bmatrix} u_{sx} \\ u_{sy} \\ u_{rx} \\ u_{ry} \end{bmatrix} =
\begin{bmatrix}
R_s + pL_s & -\omega_g L_s & pL_m & -\omega_g L_m \\
\omega_g L_s & R_s + pL_s & \omega_g L_m & pL_m \\
pL_m & -(\omega_g - \omega_r)L_m & R_r + pL_r & (-\omega_g - \omega_r)L_r \\
(\omega_g - \omega_r)L_m & pL_m & (\omega_g - \omega_r)L_r & R_r + pL_r
\end{bmatrix}
\begin{bmatrix} i_{sx} \\ i_{sy} \\ i_{rx} \\ i_{ry} \end{bmatrix},
\tag{4}
$$

Where u_{sx}, u_{sy} and i_{sx}, i_{sy} are stator voltages and currents in the general reference frame. Identical considerations remain for rotor quantities. If $\omega_g = 0$, it is obtained the 'conmutator model', but if it is employed $\omega_g = \omega_{syn}$ the expression [4] can be rewritten as:

$$
\begin{bmatrix} u_{sx} \\ u_{sy} \\ u_{rx} \\ u_{ry} \end{bmatrix} =
\begin{bmatrix}
R_s + pL_s & -\omega_1 L_s & pL_m & -\omega_1 L_m \\
\omega_1 L_s & R_s + pL_s & \omega_1 L_m & pL_m \\
pL_m & -s\omega_1 L_m & R_r + pL_r & -s\omega_1 L_r \\
s\omega_1 L_m & pL_m & s\omega_1 L_r & R_r + pL_r
\end{bmatrix}
\begin{bmatrix} i_{sx} \\ i_{sy} \\ i_{rx} \\ i_{ry} \end{bmatrix},
\tag{5}
$$

with ω_{syn} the synchronous speed and $\omega_{syn} - \omega_r = s\omega_{syn}$ with s the slip.

3. Thevenin equivalent of asynchronous machines

3.1. General reference frame - Cartesian coordinates

As presented in Appendix A and beggining with [5,13,14]:

$$
\bar{u}_{sg} = R_s \bar{i}_{sg} + j\omega_g \bar{\psi}_{sg}
\tag{6}
$$

$$
\bar{u}_{rg} = R_r \bar{i}_{rg} + \frac{d\bar{\psi}_{rg}}{dt} + j(\omega_g - \omega_r)\bar{\psi}_{rg},
\tag{7}
$$

the Thevenin equivalent of an asynchronous machine in cartesian coordinates is:

$$
\bar{u}'_{sg} = -R_s j\bar{i}_{sg} + \omega_g \left(L_s - \frac{L_m^2}{L_r} \right) \bar{i}_{sg} + \bar{u}_{sg}
\tag{8}
$$

$$\frac{d\bar{u}'_{sg}}{dt} = \omega_g \frac{L_m}{L_r} j\bar{u}_{rg} + R_r \frac{L_m^2}{L_r^2} \omega_g j\bar{i}_{sg} - (\omega_g - \omega_r) j\bar{u}'_{sg} - \frac{R_r}{L_r} \bar{u}'_{sg} \qquad (9)$$

Meanwhile expression (9) shows the internal voltage dynamics, expression (8) indicates how the stator current changes when u'_{sg} varies (assuming that $\bar{u}_{sg}$ is constant because of the induction machine is connected to an electrical grid where the voltage connection remains constant).

For clarity's sake, a typical qualitative analysis involves next steps which are carried out over a fixed speed (squirrel cage) wind generator:

1. At steady state $\frac{d\bar{\psi}_{rg}}{dt} = 0$.

2. At time $t = t_1$ there is a rotor speed change (wind velocity changed).

3. From (7) with $u_{rg} = 0$ a rotor flux variation appears $\left(\frac{d\bar{\psi}_{rg}}{dt} \neq 0\right)$.

4. Rotor flux change produces an internal stator voltage change provided that
$$\bar{u}'_{sg} = j\omega_g \frac{L_m}{L_r} \bar{\psi}_{rg}.$$

5. $\bar{u}'_{sg}$ dynamics evolves according to (9).

Active and reactive power are changed by modifying the internal voltage source. Indeed, provided that $\bar{u}_{sg} = constant$, i_{sg} must change and then the associated powers. As a consequence, in squirrel cage wind generators active and reactive powers change according to the wind velocity. i.e. they are uncontrollable from the wind generator point of view. On the other side, it is known that by modifying pitch blades the active power from squirrel cage wind generators can be regulated. In this case the reactive power is a consequence of the active power control.

3.2. Dynamic model in polar coordinates

The Thevenin equivalent, i.e. the internal voltage magnitude and its phase, allows analyzing and considering induction machines (fixed and variable speed wind generators) in an electrical perspective looking for integrating wind generators when studying stability issues of power systems.

Beginning with the Cartesian Cordinates:

$$\frac{d\bar{u}'_{sg}}{dt} = \frac{d}{dt}\sqrt{u'^2_{sx} + u'^2_{sy}},$$

$$\frac{d\bar{u}'_{sg}}{dt} = \frac{1}{\sqrt{u'^2_{sx} + u'^2_{sy}}}\left(2u'_{sx}\frac{du'_{sx}}{dt} + 2u'_{sy}\frac{du'_{sy}}{dt}\right),$$

From expressions (8) and (9) it is possible to obtain the internal voltage derivative and its phase:

$$\frac{d\bar{u}'_{sg}}{dt} = \frac{1}{\bar{u}'_{sg}}\left[u'_{sx}\left(-\omega_g\frac{L_m}{L_r}u_{ry} - R_r\frac{L_m^2}{L_r^2}\omega_g i_{sy}\right) + u'_{sy}\left(\omega_g\frac{L_m}{L_r}u_{rx} + R_r\frac{L_m^2}{L_r^2}\omega_g i_{sx}\right) - \frac{R_r}{L_r}\left(u'^2_{sx} + u'^2_{sy}\right)\right] \quad (10)$$

Because of $\tan\delta = \frac{u'_{sy}}{u'_{sx}}$, the voltage phase derivative is

$$\frac{d\delta}{dt} = \frac{1}{\bar{u}'_{sg}}\left[u'_{sx}\left(\omega_g\frac{L_m}{L_r}u_{rx} + R_r\frac{L_m^2}{L_r^2}\omega_g i_{sx}\right) - u'_{sy}\left(-\omega_g\frac{L_m}{L_r}u_{ry} - R_r\frac{L_m^2}{L_r^2}\omega_g i_{sy}\right) - (\omega_g - \omega_r)\left(u'^2_{sx} + u'^2_{sy}\right)\right] \quad (11)$$

3.3. Wind generator model considering active and reactive power delivered

By remembering classical expressions of active and reactive powers:

$$P = \frac{3}{2}Re(\bar{u}\bar{i}^*) = \frac{3}{2}(u_x i_x + u_y i_y) \quad (12)$$

$$Q = \frac{3}{2}Im(\bar{u}\bar{i}^*) = \frac{3}{2}(u_y i_x - u_x i_y), \quad (13)$$

and considering expressions (10) and (11):

$$\frac{d\bar{U}'_{sg}}{dt} = \frac{1}{\bar{U}'_{sg}}\left[\frac{2}{3}R_r\frac{L_m^2}{L_r^2}\omega_g Q'_s + u'_{sx}\left(-\omega_g\frac{L_m}{L_r}u_{ry}\right) + u'_{sy}\left(\omega_g\frac{L_m}{L_r}u_{rx}\right) - \frac{R_r}{L_r}\left(u'^2_{sx} + u'^2_{sy}\right)\right] \quad (14)$$

$$\frac{d\delta}{dt} = \frac{1}{\bar{U}'_{sg}}\left[\frac{2}{3}R_r\frac{L_m^2}{L_r^2}\omega_g P'_s + u'_{sx}(\omega_g\frac{L_m}{L_r}u_{rx}) - u'_{sy}\left(-\omega_g\frac{L_m}{L_r}u_{ry}\right) - (\omega_g - \omega_r)\left(u'^2_{sx} + u'^2_{sy}\right)\right] \quad (15)$$

it is noted, as expected, that the internal voltage derivative is a function of reactive power while its phase depends on the active power. Note also, that active and reactive powers, P'_s and Q'_s, respectively, are internal power sources, i.e. they are not the wind generator output powers.

4. Grid flux reference frame

As it was previously indicated, vector control allows controlling machine behaviors in an easier way than others techniques. One of the advantages of properly choosing the reference frame position is that is possible to simplify analysis and control[1] of electrical machines. In this way, virtual flux reference frame is chosen by considering a virtual flux from grid voltage [16,17]:

$$\bar{\psi}_g = \frac{u_s}{j\omega_g} = -j\frac{|u_s|e^{j\theta_g}}{\omega_g},$$

(16)

With θ_g the voltage phase and ω_g the phasor speed. According to Figure 3 the proposed reference frame defines a virtual flux $(\bar{\psi}_g)^\circ$ away from the voltage grid which belongs to the imaginary axis.

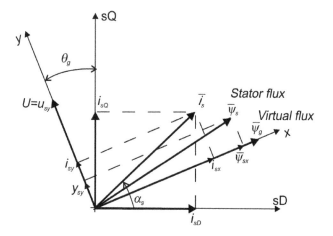

Figure 3. Stationary reference frame sD-sQ and virtual reference frame x-y

Power expressions, by virtue of the chosen reference frame, are:

$$P = \frac{3}{2}Re(\bar{u}\bar{i}^*) = \frac{3}{2}(u_x i_x + u_y i_y) = \frac{3}{2}(U i_{sy})$$

$$Q = \frac{3}{2}Im(\bar{u}\bar{i}^*) = \frac{3}{2}(u_y i_x - u_x i_y) = \frac{3}{2}(U i_{sx}),$$

---

1 Explaining vector control is not the scope of this chapter. About vector control of different kinds of machines see [12] and [15]

where both expressions show the importance of choosing suitably the reference frame. Indeed it is really simple to control active and reactive power from controlling the stator currents in an independent way.

4.1. Squirrel cage machine as a load

According to the previously discussed, it is possible to obtain active and reactive power dynamic models of the induction motors by operating with the presented expressions. These models can be employed in power systems stability studies considering that induction motors are about 60% of all loads [18]. In this way, it is better to begin with the cartesian coordinates already presented and consider a virtual flux reference frame for an induction machine motor. According to Appendix B:

$$P \cong \frac{3U^2 L_m(\omega_g - \omega_r)B}{2R_s L_s \omega_g} - \frac{3UL_m A}{2R_s L_s}\frac{dU}{dt} + \frac{3U^2 L_m A}{2R_s L_s \omega_g}\frac{d\omega_g}{dt} \cong K_{01}U^2 + K_{02}\frac{U^2}{\omega_g + \Delta\omega_g} + K_1 U\frac{dU}{dt} + K_2\frac{U^2}{\omega_g + \Delta\omega_g}\frac{d\omega_g}{dt}$$

$$Q \cong \frac{\frac{3}{2}U^2}{\omega_g L_s + L_m A\frac{d\omega_g}{dt} + L_m(\omega_g - \omega_r)B} = \frac{\frac{3}{2}U^2}{(\omega_g + \Delta\omega_g)(L_s + L_m B) + L_m A\frac{d\omega_g}{dt} - L_m\omega_r B}$$

In both power expressions $A = \left(L_s - \frac{L_m^2}{L_r}\right)\frac{L_r}{L_m R_r}$, $B = \frac{L_r}{L_m R_r}R_s$, ω_g is the line frequency, $\Delta\omega_g$ is the frequency deviation and ω_r is the rotor speed that, because of rotor inertia, remains practically constant in the temporal interval of interest.

5. Modelling of DFIG's operated with vector control

Variable speed wind farms powered by double fed induction (wound rotor) generators (DFIGs) are the other power plants considered in this chapter. Figura 4 shows the main components of a DFIG wind turbine: the rotor, the mechanical transmission system, the doubly fed induction generator and the back to back converters with their respective controls. In general, converters C1 and C2 are operated in an independent way. Meanwhile C1 is operated via vector control driving active and reactive stator powers, C2 maintains the DC bus voltage constant. In subsynchronous speeds, the rotor of the DFIG machine consumes active power meanwhile at supersynchronous speeds delivers it. As a consequence when considering active power delivered by a DFIG wind generator it should be taking into account the active power in the rotor channel.

Because of DFIG machines control is made via C1 converter, all of this chapter considers only C1 control and avoids analyzing C2 even when its operation is similar.

Beginning with the cartesian model in the general reference frame:

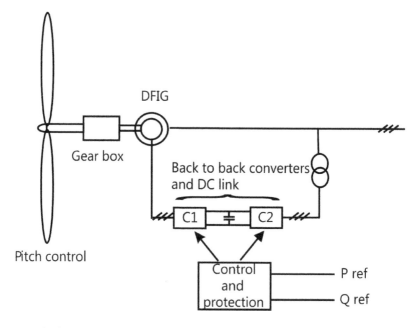

Figure 4. Wind turbine main components

$$\bar{u}'_{sg} = -R_s j \bar{i}_{sg} + \omega_g \left(L_s - \frac{L_m^2}{L_r} \right) \bar{i}_{sg} + \bar{u}_{sg}, \tag{17}$$

$$\frac{d\bar{u}'_{sg}}{dt} = \omega_g \frac{L_m}{L_r} j \bar{u}_{rg} + R_r \frac{L_m^2}{L_r^2} \omega_g j \bar{i}_{sg} - (\omega_g - \omega_r) j \bar{u}'_{sg} - \frac{R_r}{L_r} \bar{u}'_{sg}, \tag{18}$$

it is noted, due to "j" operator, that there are unwanted coupling terms between rotor and stator circuits along y and x axes. This coupling can be eliminated by utilizing an additional voltage component $\bar{u}_{rgdec} = \bar{u}_{rgdec1} + \bar{u}_{rgdec2}$ in expression (18), where:

$$\bar{u}_{rgdec1} = -R_r \frac{L_m}{L_r} \bar{i}_{sg} \tag{19}$$

$$\bar{u}_{rgdec2} = \frac{L_r}{L_m} \frac{(\omega_g - \omega_r)}{\omega_g} \bar{u}'_{sg} \tag{20}$$

Both values should be added to the rotor controllers. Note also that, according to (20)

$$\bar{u}_{rgdec2} = \frac{L_r}{L_m}\frac{(\omega_g - \omega_r)}{\omega_g}\bar{u'}_{sg} = \frac{L_r}{L_m}s\bar{u}_{sg} \cong s\bar{u'}_{sg}$$

with $L_r = L_m + L_{lr} \cong L_m$, s the slip and L_{lr} the leakage inductance. Last expression can be approximated, considering that $z = R_s + j\omega_g\left(L_s - \frac{L_m^2}{L_r}\right)$ in (17) is small. Then, $\bar{u'}_{sg} \cong U$ and $\bar{u}_{rg}$ limits the variable speed range operation in modern DFIGs by approxiimatedly 30 slip when considering rotor converter size (nominal voltage=30%U grid).

On the other side, even when the voltage feedforward can avoid the undesired coupling, this is not an optimum solution when expression (19) is seen under next considerations

• it is important to maximize the DFIG variable speed range operation, then some coupling can be tolerated,

• as presented later, looking for mimicking conventional synchronous plants operation and control, it can be useful to use the current input, eliminated by (19), for control purposes.

Figure 5 shows, on the left, a block diagram where undesired voltages are eliminated via the feedforward of the stator voltage and current; on the rigth, a simplified equivalent loop where only appears the rotor dynamics. Power references are transformed to rotor voltages ones by vector control [12] indicated as K in Figure 5.

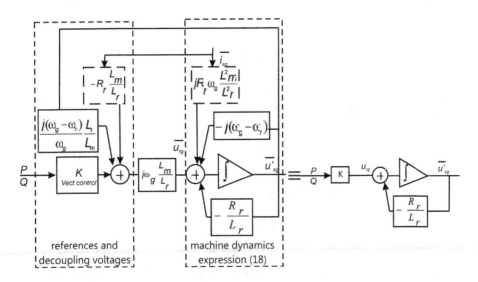

Figure 5. Feed for ward corrections and power references

According to expression (17) and considering $R_s \cong 0$, it is possible to obtain stator currents and, then, delivered powers:

$$i_{sg} = \frac{1}{\left(L_s - \dfrac{L_m^2}{L_r}\right)\omega_g}\left(\bar{u}'_{sg} - \bar{u}_{sg}\right)$$

$$P = \frac{3}{2}u_{sy}i_{sy} = \frac{3}{2}\frac{U\left(\bar{u}'_{sy} - U\right)}{\left(L_s - \dfrac{L_m^2}{L_r}\right)\omega_g} \tag{21}$$

$$Q = \frac{3}{2}u_{sy}i_{sx} = \frac{3}{2}\frac{U\bar{u}'_{sx}}{\left(L_s - \dfrac{L_m^2}{L_r}\right)\omega_g} \tag{22}$$

This completes the machine model which is presented in Figure 6 where Δ represents uncertainties and approximation errors (for example by neglecting R_s or those due to unmatched parameters between the model and the actual machine). After active and reactive powers are measured they are feedback to the controller which is usually a PI one. Note that in vector control (which is presented here by K') there are more than one feedback loop. Indeed, aside from decoupling voltages (Figure 5) there are two rotor current loops with the externals ones for active and reactive powers control in Figure 6.

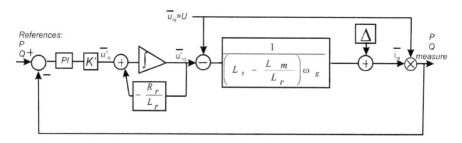

Figure 6. Decoupled model. External loop for active and reactive power control

6. Static models of asynchronous machines

It is well known that some model simplifications, keeping a good compromise between behavior and results exactitude, allow a better qualitative understanding of different kind of processes. An example of this appears studying mechanical behaviors of a wind turbine in presence of a wind velocity change. In this case, and beginning from a steady state condition, electrical behavior can be considered significantly faster than the mechanical response.

In this way, the dominant mode is the mechanical one and the electrical modes can be omitted (the stator transitory can be neglected as done in this chapter).

This section obtains the static model of induction machines derived from the dynamic one already presented. Obviously, there will always be time frames in which none of the models presented in this chapter is the best one to analyze a particular problem. In this way, if it is of importance to study the stator transitory of electrical machines, obviously, this dynamic can not be neglected as in this chapter was done. In any case, the importance of determining which model is more adequate to analyze a particular situation always lies in the physical knowledge underlying the problem.

6.1. Polar coordinates

Beginning with the dynamic model:

$$\bar{u}'_{sg} = -R_s j \bar{i}_{sg} + \omega_g \left(L_s - \frac{L_m^2}{L_r} \right) \bar{i}_{sg} + \bar{u}_{sg}$$

$$\frac{d\bar{u}'_{sg}}{dt} = \omega_g \frac{L_m}{L_r} j \bar{u}_{rg} + R_r \frac{L_m^2}{L_r^2} \omega_g j \bar{i}_{sg} - (\omega_g - \omega_r) j \bar{u}'_{sg} - \frac{R_r}{L_r} \bar{u}'_{sg} =$$

then:

$$\bar{U}'_{sg} = \sqrt{u'^2_{sx} + u'^2_{sy}} \text{ and } \delta = \arctan\left(\frac{u'_{sy}}{u'_{sx}} \right)$$

6.2. Static model as a function of the active and reactive stator powers delivered in the virtual flux reference frame

The voltage internal phase:

$$\tan\delta = \frac{u'_{sy}}{u'_{sx}}$$

$$\tan\delta = \frac{-R_s Q + \omega_g \left(L_s - \frac{L_m^2}{L_r} \right) P + \frac{3}{2} U^2}{R_s P + \omega_g \left(L_s - \frac{L_m^2}{L_r} \right) Q}$$

is highly dependent on grid voltage, on active power and, in a lesser way, on reactive power. Note that, if a DFIG is operated at unitary power factory, i.e. $Q = 0$, then $\delta \cong 90°$.

On the other side, the internal voltage resuls:

$$\bar{U}'_{sg} = \sqrt{u'^2_{sx} + u'^2_{sy}} = u'_{sy}\sqrt{\cot\delta^2 + 1} = \frac{u'_{sy}}{\sin\delta} \Rightarrow$$

$$\overline{u}'_{sg} = \frac{-R_s\frac{2}{3}Q + \omega_g\left(L_s - \frac{L_m^2}{L_r}\right)\frac{2}{3}P + U^2}{U\sin\delta}$$

$$\overline{u}'_{sg} = \sqrt{u'^2_{sx} + u'^2_{sy}} = u'_{sx}\sqrt{\tan\delta^2 + 1} = \frac{u'_{sx}}{\cos\delta} \Rightarrow$$

$$\overline{u}'_{sg} = \frac{R_s\frac{2}{3}P + \omega_g\left(L_s - \frac{L_m^2}{L_r}\right)\frac{2}{3}Q}{U\cos\delta}$$

7. Wind farm control

Wind farms have become a visible component of interconnected power grids. In the beginnings of wind generation, when a low portion of the electrical power was delivered from this renewable energy, only simple engineering judgment were necessary to conclude about the negligible impact of wind generation on power systems. Nowadays, with wind farms, but also with high power wind generators in dispersed grids, approaching the output rating of conventional power plants, it is necessary understanding the way in which wind generation can impact and/or contribute to the power system stability.

This section presents different wind farms controls by considering them as current sources and voltages sources (Thevenin equivalent) in an effort for mimicking conventional generation.

The proposed wind farm control approaches are based in the named Lyapunov Theory which give place to linear and non linear wind farm controls. In these approaches, the DFIG capability for controlling active and reactive powers plays an important role in contributing to power system stability. Then, from here on, only DFIG wind farms are considered.

Additionally, Energy (Lyapunov) approach is not based on system linearization and the proposed analysis allows considering any wind farm in any power system in the same way avoiding transform every issue about integrating wind farms in power systems in a different problem.

As indicated, in order to contribute to the network stability, both active and reactive wind farm power controls of a wind farm are considered. Then, steady state controls (normal operating conditions) plus incremental corrections are proposed:

$$Q_{wf} = Q_{SC} + \Delta Q,$$

$$P_{wf} = P_{SC} + \Delta P.$$

with $P_{wf}(Q_{wf})$ is the total active (reactive) power, $P_{SC}(Q_{SC})$ the power reference given by a Supervisory Control [19,20] and $\Delta P(\Delta Q)$ the wind farm correction which contributes to the power system stability.

Note that for both active and reactive corrective actions some power reserve is required. Indeed, about active power it will be expected a power reserve which is a function of the wind turbine operating point and about the reactive power correction, the total 'apparent' power of the DFIG machine will limit the corrective action.

7.1. Wind farm aggregated model

A complete model of a wind farm with a high number of wind generators, may lead to compute an excessive and impractical number of equations. The size of the wind farm model may be reduced by aggregating several wind turbines with similar incoming wind into a bigger turbine called aggregated turbine [14]. The mechanical and electric parameters per unit are preserved, and the nominal power is increased up to the sum of the nominal power of the whole set of turbines to obtain the parameters of the aggregated turbine. This procedure is employed in this chapter where the wind farm is modeled as one aggregated wind turbine. In this regard, the Thevenin equivalent can be obtained as in a classical problem of electrical systems, applying precisely Thevenin theorem, being the internal voltage and its phase calculated according to the impedance and voltage seen from the wind farm common connection point.

7.2. Wind farm control. Method of Lyapunov

Power system stability has been defined as that property of a power system that enables it to remain in a state of operating equilibrium under normal operating conditions and to regain an acceptable state of equilibrium after being subjected to a disturbance [5].

Lyapunov demonstrated that a nonlinear dynamic system:

$$\dot{x} = f(x), \quad f(0) = 0, \tag{23}$$

around the equilibrium point $x=0$ is asymptotically stable if there exist a scalar function $v(x)>0$ for (23) with derivative $\dot{v}(x)<0$. The last condition is relaxed to $\dot{v}(x)\leq00$ provided that $\dot{v}(x)=0$ only vanish at $x=0$.

Lyapunov theory deals with dynamical systems without inputs. However, it is possible to employ Lyapunov theory in feedback design by making negative the Lyapunov derivative [21,22,23]. The incremental energy function v of a power system without wind farms, where conventional models of the synchronous generators and of load impedance are considered, is [23]:

$$v = \sum_{k=1}^{N_G}(0.5M_k\tilde{\omega}_k^2 - P_{Mk}\tilde{\delta}_k) + \sum_{k=1}^{N_L}(P_{Lk}\tilde{\theta}_k + \int \frac{Q_{Lk}}{V_k}dV_k), \tag{24}$$

with

$$\widetilde{\omega}_k = \omega_k - \omega_{COI}, \qquad \omega_{COI} \doteq \frac{1}{M_T} \Sigma_{k=}^{m} M_k \omega_k,$$

$$\delta_k = \delta_k - \delta_{COI}, \qquad \delta_{COI} \doteq \frac{1}{M_T} \Sigma_{k=}^{m} M_k \delta_k,$$

$$\theta_k = \theta_k - \delta_{COI}, \qquad M_T \doteq \Sigma_{k=}^{m} M_k,$$

with N_L and N_G the number of loads and generators, M_k the machine inertia constant, ω_k the machine speed, δ_k the angle of the voltage behind the transient reactance, θ_k the angle at each bus, P_{M_k} the mechanical power of the generators, P_{Lk} and Q_{Lk} the active and reactive load powers and V_k the voltage at the connection point. The angles and speeds are measured with respect to the center of inertia (COI) reference frame (δ_{COI} and ω_{COI}).

7.2.1. Non-linear power control of wind farms as negative loads

To damp the electromechanical oscillations, i.e. frequency oscillations, the incremental energy function of the power system must decrease. The time derivative of this energy function considering wind farms as negative loads, i.e. acting through its active and reactive powers, yields:

$$\dot{\nu} = \sum_{k=1}^{N_G} (M_k \dot{\widetilde{\omega}}_k^2 + P_{Gk} - P_{Mk}) \dot{\widetilde{\delta}}_k + \sum_{n=1}^{N_L} P_n \dot{\widetilde{\theta}}_n + \sum_{n=1}^{N_L} \frac{\dot{V}_n}{V_n} Q_n - \frac{\dot{V}_{wf}}{V_{wf}} \Delta Q - \Delta P \dot{\widetilde{\theta}}_{wf}. \qquad (25)$$

Then, the active and reactive powers (ΔP and ΔQ) must be chosen in order to allow the sufficient condition of the derivative of the incremental energy function. In equation (25), the expression in between parenthesis is zero (or lesser than zero) because of the generators power balance (equals the internal generator damping). The next two terms correspond to the power balance equations at the nodes and are zero. Then, looking for damping the electromechanical oscillations, the last two expressions must be less than zero.

About the active power two possibilities are choosen [10]:

$$\Delta P = K_{c1} \dot{\widetilde{\theta}}_{wf}; \quad with \quad K_{c1} > 0, \qquad (26)$$

$$\Delta P = K_{c2} \ddot{\widetilde{\theta}}_{wf}^2 \dot{\widetilde{\theta}}_{wf}; \quad with \quad K_{c2} > 0. \qquad (27)$$

meanwhile the first expression which is the classical proportional frequency law, the second one is a kind of inertial response. Indeed, the second expression can be understood as a modification of (26) where K_{c1} is a variable gain $K_{c1} = K_{c2} \ddot{\theta}_{wf}^2$ which takes into account frequency derivative.

On the other hand, a non-linear control strategy of the wind farm through the reactive power ΔQ can be derived from expression (25) by considering that the wind farm emulates the behavior of a static var compensator:

$$\Delta Q = b_u V_{wf}^2 \quad \Rightarrow$$

$$-\frac{\dot{V}_{wf}}{V_{wf}} \Delta Q = -b_u V_{wf} \dot{V}_{wf} = -\frac{1}{2} b_u \frac{d}{dt} V_{wf}^2.$$

In this way, in order to keep the sufficient condition, some possibilities arise:

$$b_u = K_r \frac{dV_{wf}^2}{dt}; \quad \text{with} \quad K_r > 0 \quad or \tag{28}$$

$$b_u = K_r sign\left(\frac{dV_{wf}^2}{dt}\right); \quad \text{with} \quad K_r > 0, \tag{29}$$

Where b_u is the (equivalent) wind farm susceptance.

With respect to the active power control, note that the idea of a power reserve, as a percentage of the maximum available power, is very attractive from a point of view of the network stability and it is usually employed [19,20]. However, suppose that $\Delta P = 0$, the appropriate choice of the reactive power function (expressions (28) or (29)) implies that the energy function derivative (34) almost always decreases, i.e. any electromechanical oscillation is damped. Then, the wind farm reactive power contributes to damp the electromechanical oscillations of the power system. Being, in general, $\Delta P \neq 0$, the reactive power function reinforces the wind farm contribution to the network stability.

Note that, with both active and reactive control laws, it is possible maximizing the use of the energy resource in order to contribute to damp the electromechanical oscillations by exploiting all the capabilities of the DFIG machines. In this way, it is possible designing a control law for the reactive power which takes into account the advantage of producing as much reactive power as possible considering the apparent power of the DFIGs.

7.2.2. Non-linear active power control of wind farms as Thevenin equivalent

Due to it is expected that wind farms act as power plants [24], it is necessary to demostrate that wind farms behave as their equivalent synchronous generators (the conventional power plants) with proportional and derivative (inertial) frequency control laws. The equations representing the dynamic behavior of synchronous generators for the reduced model, are [5]:

$$\dot{\tilde{\delta}}_k = \tilde{\omega}_k, \tag{30}$$

$$\dot{\tilde{\omega}}_k = \frac{1}{M_k}(P_{mk} - D_k \omega_k - P_{Gk}). \tag{31}$$

where M_k is the inertia of the whole machine (synchronous generator plus prime mover), P_{mk} is the mechanical power produced by the prime mover, D_k is the component of internal friction of the generator and P_{Gk} is the electrical power injected in the network. Because of variables are in per unit, powers and torques are equal [5]. Mimicking the analysis for a wind farm with frequency control and inertial contribution yields [19]:

$$\dot{\tilde{\alpha}}_k = \Omega_{gk}, \tag{32}$$

$$\dot{\tilde{\Omega}}_{gk} = \frac{1}{M_{gk}}(P_t - P_{wf}), \tag{33}$$

with P_t the turbine power.

Considering the control as:

$$P_{wf} = P_{SC} + \Delta P,$$

$$\Delta P = K_p(\tilde{\omega}_{ref} - \tilde{\omega}) + K_d(\dot{\tilde{\omega}}_{ref} - \dot{\tilde{\omega}}) = K_p\breve{\omega} + K_d\dot{\breve{\omega}},$$

where is included a PD control named "proportional and inertial" classical control laws, then:

$$\dot{\tilde{\omega}} = \frac{1}{K_d}(P_t - (K_p\tilde{\omega} + M\dot{\tilde{\Omega}}_{gk}) - P_{SC}). \tag{34}$$

Note the similarity of this expression with (31) for synchronous generation. Then, the analysis focuses on addressing the control of the wind farm as in $\dot{v}_1$ which considers the derivative of the Energy Function of a synchronous generator in (34):

$$v_1 = \sum_{k=}^{N_G}(M_k\tilde{\omega}_k^2 + P_{Gk}\theta_k - P_{Mk}\theta_k) \quad \Rightarrow \dot{v}_1 = \sum_{k=}^{N_G}(-D\tilde{\omega}_k) < \tag{35}$$

The equivalent expression for a wind farm is:

$$v_{1wf} = \breve{\omega}_k^2 + \frac{P_{SC}}{K_d}\breve{\theta}_{wf} - \frac{P_t}{K_d}\breve{\theta}_{wf} \quad \Rightarrow \dot{v}_{1wf} = \breve{\omega}_k\breve{\omega}_k + \frac{P_{SC} - P_t}{K_d}\dot{\breve{\theta}}_{wf} = (\breve{\omega}_k + \frac{P_{SC} - P_t}{K_d})\dot{\breve{\theta}}_{wf},$$

$$\dot{v}_{1wf} = \left(-\frac{M_{wf}}{K_d}\dot{\Omega}_g - \frac{K_p}{K_d}\tilde{\omega}\right)\dot{\breve{\theta}}_{wf} = -\frac{M_{wf}}{K_d}\dot{\Omega}_g\dot{\breve{\theta}}_{wf} - \frac{K_p}{K_d}\dot{\breve{\theta}}_{wf}^2, \tag{36}$$

being $K_p>0$ and $K_d>0$ it is verified the negative sign of the second term in the last expression. On the other side, in order to verify the negative sign of $-\dfrac{M_{wf}}{K_d}\dot{\Omega}_g\overset{\smile}{\theta}_{wf}$ it is necessary to analyse the wind farm convergence to an equilibrium point (e.p.), knowing that $M_{wf}>$ and considering that the wind farm (the aggregated turbine) is outside the equilibrium point.

Figure 7 presents the torque - speed curves of the aggregated wind turbine with the wind velocity as a parameter. The e.p., considering constant wind velocity, corresponds to nominal frequency at the wind farm connection point with constant shaft speed $\Omega_{e.p.}$ of the aggregated turbine.

In order to verify the convergence to the equilibrium, two conditions outside the e.p. [19], which are consequence of electrical disturbances, will be analyzed. First, consider that because of a disturbance action, the aggregated wind turbine is operating at point A (Figure 7) with $\Omega_A<\Omega_{e.p.}$, being the frequency $\overset{\smile}{\theta}_A<\overset{\smile}{\theta}_{e.p.}$. At that point, the wind farm generated power is higher than the nominal one, i.e. the wind farm is contributing to restore the frequency at the connection point.

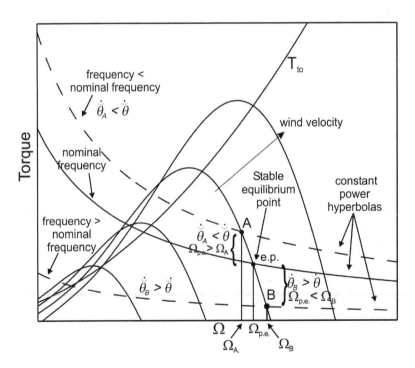

Figure 7. Torque - speed characteristics. Convergent behaviour to the e.p.

When the disturbance disappears, the network returns to its normal configuration and the wind farm power is higher than that which maintains the power balance in the system. As the deviation of the frequency decreases, so does the wind farm generated power. As a consequence, the wind turbine torque decreases and the turbine speed experiences an increment until the speed reaches $\Omega_{p.e.}$. Thus, while the frequency reaches their nominal value, the wind turbine increasing its speed. Then, the (negative) sign of (36) is verified by considering:

$$\dot{\theta}_{wf} = \dot{\theta}_{ref} - \dot{\theta}_A > 0 \quad and \tag{37}$$

$$\dot{\Omega}_g = \frac{\Omega_{e.p.} - \Omega_A}{\Delta t} > 0. \tag{38}$$

If as a consequence of another disturbance the aggregated turbine is operating at point B (Figure 7), when the disturbance is removed the sign of (36) yields:

$$\dot{\theta}_{wf} = \dot{\theta}_{ref} - \dot{\theta}_A < 0 \quad and \tag{39}$$

$$\dot{\Omega}_g = \frac{\Omega_{e.p.} - \Omega_B}{\Delta t} < 0. \tag{40}$$

Expressions (38) and (40) verify the negative sign of (36).

7.2.3. Non-linear reactive power control of wind farms from the Thevenin equivalent

In last subsection, by mimicking the reduced model of conventional synchronous generators with wind farms, it has been deduced that wind farms PD laws, i.e. the named frequency and inertial responses control approaches, allows to contribute to the power system stability. In order to deduce a wind farm reactive power control, it is necessary to include the named "Structure Preserving the Model" [21] or "the one axis model" [6] for the conventional synchronous generators. The dynamics of a $k-th$ synchronous generator, respect to the COI is [21]:

$$\dot{\delta}_k = \tilde{\omega}_k, \tag{41}$$

$$M_k \dot{\tilde{\omega}}_k = P_{mk} - D_k \omega_k - P_{Gk} - \frac{M_k}{M_T} P_{COI} \tag{42}$$

$$T'_{dok} \frac{dE'_{qk}}{dt} = \frac{x_{dk} - x'_{dk}}{x'_{dk}} V_{n+k} \cos(\delta_k - \theta_{n+k}) + E_{fdk} - \frac{x_{dk}}{x'_{dk}} E'_{qk} \tag{43}$$

being T'_{dok} the d axis transient open circuit time constant; E'_{qk} the q axis voltage behind transient reactance, E_{fdk} is the exciter voltage which is assumed constant (if the exciter control action is included in the generator model, at least one additional dynamic expression should be included) and x_{dk} and x'_{dk} are d axis synchronous reactance and transient reactance, respectively.

An advantage of this model when compared to the classical one with two states (expressions (30) and (31)) is the possibility of including loads where the impedance are not constant. In order of making more clear the explanation it will not be included any wind farm in this step. According to [21] next terms are added to the already applied Lyapunov function:

$$V_{2a} = \sum_{i=n+1}^{2n} \frac{1}{2x'_{di-n}} [E'^{2}_{qi-n} + V_i^2 - 2E'_{qi-n} V_i \cos(\delta_{i-n} - \theta_i)]$$

$$V_{2b} = -\sum_{i=}^{n} \frac{E_{fdi} E'_{qi}}{x_{di} - x'_{di}}$$

$$V_{2c} = \sum_{i=}^{n} \frac{E'^{2}_{qi}}{(x_{di} - x'_{di})}$$

$$V_{2d} = -\frac{1}{2} \sum_{i=n+1}^{n+N} \sum_{l=n+1}^{n+N} B_{kl} V_k V_l \cos(\theta_i - \theta_l)$$

$$V_{2e} = \sum_{i=n+1}^{2n} \frac{x'_{di-n} - x_{di-n}}{4x'_{di-n} x_{di-n}} [V_i^2 - V_i^2 \cos((\delta_{i-n} - \theta_i))]$$

Then, the derivative of the (new) Lyapunov function is:

$$\dot{v} = \sum_{k=}^{N_G} (M_k \tilde{\omega}_k^2 + P_{Gk} - P_{Mk}) \dot{\delta}_k + \sum_{n=}^{N_L} P_n \dot{\theta}_n + \sum_{n=}^{N_L} \frac{\dot{V}_n}{V_n} Q_n + \dot{v}_{2a} + \dot{v}_{2b} + \dot{v}_{2c} + \dot{v}_{2d} + \dot{v}_{2e},$$

$$\dot{v} = -\sum_{k=}^{N_G} D_k(\tilde{\omega}_k^2) - \sum_{k=}^{N_G} \frac{T'_d}{x_d - x'_d} \dot{E}'^2_q,$$

where:

$$\sum \left(P_n \dot{\theta}_n + \frac{\partial v_{2a}}{\partial \theta_n} \dot{\theta}_n + \frac{\partial v_{2d}}{\partial \theta_n} \dot{\theta}_n + \frac{\partial v_{2d}}{\partial \theta_n} \dot{\theta}_n \right) = \sum (P_n + P_k) \dot{\theta}_n = 0$$

$$\sum \left(\frac{\dot{V}_n}{V_n} Q_n + \frac{\partial v_{2a}}{\partial V_n} \dot{V}_n + \frac{\partial v_{2d}}{\partial V_n} \dot{V}_n + \frac{\partial v_{2d}}{\partial V_n} \dot{V}_n \right) = \sum (Q_n + Q_k) \frac{\dot{V}_n}{V_n} = 0$$

i.e. the total active and reactive powers (consumed plus injected) is zero.

When it is considered the last state variable, the voltage behind transient reactance E'_{qk}, in the derivative of the Lyapunov function:

$$\dot{v}_3 = \frac{\partial v_{2a}}{\partial E'_q}\frac{dE'_q}{dt} + \frac{\partial v_{2b}}{\partial E'_q}\frac{dE'_q}{dt} + \frac{\partial v_{2c}}{\partial E'_q}\frac{dE'_q}{dt} = -\sum_{k=}^{N_G}\frac{T'_d}{x_d - x'_d}\dot{E}'^2_q$$

Due to $T'_d > 0$ and $x_d > x'_d$ the Lyapunov condition is met.

Mimicking the third state equation of the Structure Preserving Model with the aggregated wind turbine (with DFIGs) implies taking dynamical expressions from (18) and eliminating an undesired coupling:

$$\frac{d\bar{u}'_{sg}}{dt} = \omega_g\frac{L_m}{L_r}j\bar{u}_{rg} + R_r\frac{L_m^2}{L_r^2}\omega_g j\bar{i}_{sg} - (\omega_g - \omega_r)j\bar{u}'_{sg} - \frac{R_r}{L_r}\bar{u}'_{sg} \tag{44}$$

$$T_r\frac{d\bar{u}'_{sg}}{dt} = \frac{L_m^2}{L_r}\omega_g j\bar{i}_{sg} + \omega_g\frac{L_m}{R_r}j\bar{u}_{rg} - \bar{u}'_{sg} \tag{45}$$

According to virtual flux control, i_{sy} controls active power meanwhile i_{sx} can be used for reactive power control $\left(Q = \frac{3}{2}u_{sy}i_{sx}\right)$. Operating with expression (45):

$$T_r\frac{du'_{sy}}{dt} = \frac{L_m^2}{L_r}\omega_g i_{sx} + \omega_g\frac{L_m}{R_r}u_{rx} - u'_{sy}$$

$$T_r\frac{du'_{sy}}{dt} = (L_s - L_{eq})\omega_g i_{sx} + \omega_g\frac{L_m}{R_r}u_{rx} - u'_{sy}$$

and considering, expression (8), that $L_{eq} = \left(L_s - \frac{L_m^2}{L_r}\right)$.

According to Figure 8, which presents the Thevenin equivalent and the correspoding phasor diagram, and considering expression (17) with $R_s \cong 0$, results:

$$\dot{i}_{sx} = \frac{1}{L_{eq}\omega_g}\left(\bar{u}'_{sx} - \bar{u}_{sx}\right) = \frac{\bar{u}'_{sx}}{L_{eq}\omega_g} = \frac{V\sin(\delta - \theta)}{L_{eq}\omega_g}$$

$$\dot{i}_{sx} = \frac{V\cos[90 - (\delta - \theta)]}{L_{eq}\omega_g}$$

Then,

$$T_r\frac{du'_{sy}}{dt} = \frac{(L_s - L_{eq})}{L_{eq}}V\cos[90 - (\delta - \theta)] + \omega_g\frac{L_m}{R_r}u_{rx} - u'_{sy},$$

$$\frac{1}{\omega_g}\frac{R_r}{L_m}T_r\frac{du'_{sy}}{dt} = \frac{1}{\omega_g}\frac{R_r}{L_m}\frac{(L_s - L_{eq})}{L_{eq}}V\cos[90-(\delta-\theta)] + u_{rx} - \frac{1}{\omega_g}\frac{R_r}{L_m}u'_{sy},$$

with δ the internal voltage phase, θ the voltage angle at the common connection point and $T_r = L_r / R_r$. Last expression is pretty similar to (43). Then, an equivalent Lyapunov analysis can be done which implies that it is possible to contribute to the power system stability as conventional generation does.

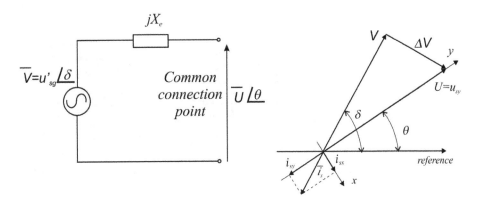

Figure 8. Wind farm Thevenin equivalent and phasor diagram

8. Tendencies in the study of wind farm contributions to system stability

It is worth to note that the objective of this chapter is not give a full list of new trends in wind farms control. The comments are focused on aspects in which, in the authors opinion, are expected some improves in the near future. This does not mean to avoid exploring other control techniques that have been successful in different fields of control systems.

In this way exploring another control Lyapunov functions will certainly be an important input not just for wind farm contributions in power systems but also for power systems stability in general. Indeed, acting as negative loads implied than wind farm control can be extrapolated to other kind of devices which can control active and/or reactive power independently, as FACTS. Other important subject is concerning with the load characterization. Then, including in the Lyapunov frame some loads models as the developed for the induction motor will allow to find new rules with the advantages of decentralized and local measures as the developed ones in the chapter.

However, there would be particular cases in which some rules based in local measures, as the frequency response, could be ineffective. This is presented in Figure 9 taking the topology of a power system from [5] and including a wind farm in different places. In part (b) of Figure 9, after a disturbance appears and because of the wind farm is in the middle of areas, wind farm frequency remains constant due to that measure is taken in the center of 'bouncing' between areas. Indeed, in Figure 9(b) wind farm is located in the COI of the system, then $\widetilde{\omega}_k = \omega_k - \omega_{COI} = \omega_{COI} - \omega_{COI} = 0$

Note that the power system in Figure 9 can be analyzed as an academic subject of study due to actual symmetrical systems will always have differences which will be enough to allow that classical control laws contribute to the power system stability. In this cases, obviously, the Lyapunov control laws derived in this chapter could be less efective than other ones which take into account the nature of this electrical grid.

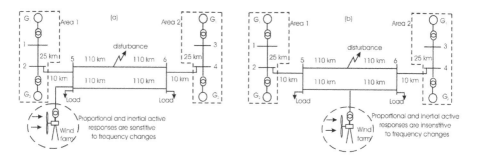

Figure 9. Example considering different wind farm locations with only active power correction based on frequency measures. (a) Effective wind farm correction from proportional and derivative (inertial) actions; (b) Ineffective wind farm active correction

About grids with symmetrical topologies, Passivity theory [25,26,27] can allow to consider some other tools looking for contributing to the power system stability [28,29]. Because of Passivity considers finding a controller in such a way the dynamical system energy function takes the desired form ('energy shaping') and lately considering a power shaping approach, but also solving an energy function which qualifies as a Lyapunov function, Passivity theory implies an important step ahead in the power systems study. Some authors attempts about including wind farm control in the passivity frame can be found in [30,31].

9. Conclusions

This chapter can be divided in two parts, meanwhile the first one is devoted of exploring models and developing a Thevenin equivalent of squirrel cage induction generators (fixed wind generators) and DFIG ones (variable speed generators), after introducing an aggregated wind farm concept and considering a wind farm as another Thevenin equivalent, the second part of the chapter considers analyzing DFIG wind farm linear and non linear control.

The final objective of the chapter is to demonstrate that can be developed an equivalent behavior of wind farms to their counterparts, the synchronous conventional generators, by properly controlling wind farms. This point of view will help to promote wind farm integration but, at the same time, it will open new doors on contributing to the power systems stability via wind farms control.

About the first part, induction machines models were presented including an additional dynamic model of a squirrel cage induction machine as a load. Additionally, even when vector control was not explained, a control perspective was adopted in showing the way in that stator currents (and powers) can be controlled from rotor voltages in DFIG machines. Also, by considering active and reactive powers delivered for wind generators as input, a static model of induction machines was derived from its dynamical counterpart.

About controlling wind farms, it was possible maximizing the use of the energy resource in order to contribute to damp the electromechanical oscillations by exploiting all the capabilities of the DFIG machines. In this way, wind farms were operated under a Supervisory Control which imposed steady states power references and (added) corrective actions, under Lyapunov Theory which was shortly explained, were proposed as a complement of the references ones.

By considering wind farms as negative loads the Lyapunov frame showed that meanwhile wind farm proportional active power control law was the same as the classical proposed one, about the inertial effect a modification from that one was found. About reactive power, wind farms emulated the behavior of static VAR compensators giving place to a highly non linear control law.

By considering wind farms as voltage sources the Lyapunov frame demonstrated that classical active power laws, proportional and derivative with frequency, can contribute to the power system stability. Also, about reactive power, it was possible demonstrating that wind farms can contribute to damping electromechanical oscillations by controlling its internal (Thevenin) voltage via rotor voltage actions in DFIGs.

As a consequence of the Energy (Lyapunov) approach, the obtained control laws were not based on the linearization of the system. This assures a bigger domain of attraction of the wind farm contribution indicating that, even under severe disturbances, the proposed control laws will contribute to the power system stability.

It is important to note the local nature of the signals used for control laws which avoids any coordination with the rest of the system. Additionally, the damping is not dependent on the power flow direction and neither the kind of failures on the power system.

However, aside from Energy Functions included only some kind of loads and that line resistances must be neglected in the calculus of the functions, another limitation can be found from considering some power systems and wind farms locations where active (or reactive) power control laws can not be effective on contributing to the system stability.

Looking for solving some of the aforementioned drawbacks, future research in Lyapunov topics were indicated but also other perspective, which in fact calculates Energy Functions in control systems, as Passivity was proposed. In this way, the same border conditions remain for future control laws as the necessity of local uncoordinated measures, independency of power system topology, robustness, etc.

Finally, it is important to note some differences faced when looking for control laws in power systems. Electrical grids have distinctive characteristics when compared with other "conventional" control systems. In first place, as indicated, the necessity and importance of uncoordinated and robust control laws which contribute to the whole system stability but based from local measures. In second place, wind farm control laws must cover an important range of applications by considering different power systems and different places in which a wind farm can be connected. Indeed, control systems teach us that it is of fundamental importance knowing the "system model", however that model is highly changing and it has infinite equilibrium points when talking about power systems. All of these characteristics only emphasize that power systems are difficult to control and as technology evolves they are upgraded in such a way that some challenges are solved and, at the same time, new ones appear.

Appendix A

Beginning with the expressions in the general reference frame and neglecting stator dynamics $\left(d\bar{\psi}_{sg}/dt = 0\right)$ [5,13,14]:

$$\bar{u}_{sg} = R_s \bar{i}_{sg} + j\omega_g \bar{\psi}_{sg} \tag{46}$$

$$\bar{u}_{rg} = R_r \bar{i}_{rg} + \frac{d\bar{\psi}_{rg}}{dt} + j(\omega_g - \omega_r)\bar{\psi}_{rg}, \tag{47}$$

with

$$\bar{\psi}_{sg} = L_s \bar{i}_{sg} + L_m \bar{i}_{rg} ; \bar{\psi}_{rg} = L_r \bar{i}_{rg} + L_m \bar{i}_{sg} \Rightarrow$$

$$\bar{i}_{rg} = \frac{\bar{\psi}_{rg} - L_m \bar{i}_{sg}}{L_r} \; and \; \bar{\psi}_{sg} = \left(L_s - \frac{L_m^2}{L_r} \right) \bar{i}_{sg} + \frac{L_m}{L_r} \bar{\psi}_{rg}$$

which replaced in expression (46) gives

$$\bar{u}_{sg} = \left[R_s + j\omega_g \left(L_s - \frac{L_m^2}{L_r} \right) \right] \bar{i}_{sg} + j\omega_g \frac{L_m}{L_r} \bar{\psi}_{rg} \Rightarrow \bar{u}_{sg} = \left[R_s + j\omega_g \left(L_s - \frac{L_m^2}{L_r} \right) \right] \bar{i}_{sg} + \bar{u}'_{sg} \Rightarrow$$

$$\bar{u}'_{sg} = -\left[R_s + j\omega_g \left(L_s - \frac{L_m^2}{L_r} \right) \right] j\bar{i}_{sg} + \bar{u}_{sg} \tag{48}$$

where $\bar{u}'_{sg}$ is the internal voltage source and the associated impedance is $\left[R_s + j\omega_g \left(L_s - \dfrac{L_m^2}{L_r} \right) \right]$. Also, from expression (47):

$$\bar{u}_{rg} = R_r \bar{i}_{rg} + \frac{d\bar{\psi}_{rg}}{dt} + j(\omega_g - \omega_r)\bar{\psi}_{rg} = R_r \frac{\bar{\psi}_{rg} - L_m \bar{i}_{sg}}{L_r} + \frac{d\bar{\psi}_{rg}}{dt} + j(\omega_g - \omega_r)\bar{\psi}_{rg},$$

$$\frac{d\bar{\psi}_{rg}}{dt} \cdot j\omega_g \frac{L_m}{L_r} = \left[\bar{u}_{rg} + R_r \frac{L_m}{L_r} \bar{i}_{sg} - R_r \frac{\bar{\psi}_{rg}}{L_r} - j(\omega_g - \omega_r)\bar{\psi}_{rg} \right] \cdot j\omega_g \frac{L_m}{L_r}$$

$$\frac{d\bar{u}'_{sg}}{dt} = \omega_g \frac{L_m}{L_r} j\bar{u}_{rg} + R_r \frac{L_m^2}{L_r^2} \omega_g j\bar{i}_{sg} - (\omega_g - \omega_r) j\bar{u}'_{sg} - \frac{R_r}{L_r} \bar{u}'_{sg} \tag{49}$$

Summarizing:

$$\bar{u}'_{sg} = -R_s j\bar{i}_{sg} + \omega_g \left(L_s - \frac{L_m^2}{L_r} \right) \bar{i}_{sg} + \bar{u}_{sg} \tag{50}$$

$$\frac{d\bar{u}'_{sg}}{dt} = \omega_g \frac{L_m}{L_r} j\bar{u}_{rg} + R_r \frac{L_m^2}{L_r^2} \omega_g j\bar{i}_{sg} - (\omega_g - \omega_r) j\bar{u}'_{sg} - \frac{R_r}{L_r} \bar{u}'_{sg} \tag{51}$$

Appendix B

$$P = \frac{3}{2}(u_{sy} i_{sy}) Q = \frac{3}{2}(u_{sy} i_{sx}) \tag{52}$$

Taking $\bar{i}_{sg}$ and $\bar{i}_{rg}$ from (46) and (47), respectively and considering $u_{sy} = U$:

$$\bar{i}_{sg} = \frac{\bar{u}_{sg} - j\omega_g \bar{\psi}_{sg}}{R_s} \text{ and } \bar{\psi}_{sg} = L_s \bar{i}_{sg} + L_m \bar{i}_{rg} \Rightarrow \bar{i}_{sg} = \frac{\bar{u}_{sg} - j\omega_g (L_s \bar{i}_{sg} + L_m \bar{i}_{rg})}{R_s},$$

$$\text{and } \bar{\psi}_{rg} = L_r \bar{i}_{rg} + L_m \bar{i}_{sg} \Rightarrow \bar{i}_{rg} = \frac{-\dfrac{d\bar{\psi}_{rg}}{dt} - j(\omega_g - \omega_r)\bar{\psi}_{rg}}{R_r}.$$

Taking the last three expressions it is possible to get $\bar{i}_{rg}$ as a function of $\bar{i}_{sg}$, $\bar{u}_{sg}$, $\dfrac{d\omega_g}{dt}$, ω_g, ω_r. Then, replacing $\bar{i}_{rg}$ in the second expression and from the first one it is possible to obtain i_{sg} as a function of $\dfrac{d\omega_g}{dt}$, $\dfrac{d\bar{u}_{sg}}{dt}$, $\bar{u}_{sg}$. Finally, active and reactive powers become:

$$P = \frac{U^2(R_s - L_m\omega_g(\omega_g - \omega_r)A)}{(R_s - L_m\omega_g(\omega_g - \omega_r)A)^2 + \left(\omega_g L_s + L_m A\dfrac{d\omega_g}{dt} + L_m(\omega_g - \omega_r)B\right)^2} -$$

$$- \frac{U\left(\omega_g L_s + L_m A\dfrac{d\omega_g}{dt} + L_m(\omega_g - \omega_r)B\right)\omega_g L_m\left(\dfrac{A}{R_s}\dfrac{dU}{dt} - \left(\dfrac{A}{R_s}\dfrac{d\omega_g}{dt} + (\omega_g - \omega_r)\dfrac{B}{R_s}\right)\dfrac{U}{\omega_g}\right)}{(R_s - L_m\omega_g(\omega_g - \omega_r)A)^2 + \left(\omega_g L_s + L_m A\dfrac{d\omega_g}{dt} + L_m(\omega_g - \omega_r)B\right)^2}$$

$$Q = \frac{U\omega_g L_m\left(\dfrac{A}{R_s}\dfrac{dU}{dt} - \left(\dfrac{A}{R_s}\dfrac{d\omega_g}{dt} + (\omega_g - \omega_r)\dfrac{B}{R_s}\right)\dfrac{U}{\omega_g}\right)}{(R_s - \omega_g L_m(\omega_g - \omega_r)A)} +$$

$$+ \frac{U^2\left(\omega_g L_s + L_m A\dfrac{d\omega_g}{dt} + L_m(\omega_g - \omega_r)B\right)}{(R_s - L_m\omega_g(\omega_g - \omega_r)A)^2 + \left(\omega_g L_s + L_m A\dfrac{d\omega_g}{dt} + L_m(\omega_g - \omega_r)B\right)^2} -$$

$$- \frac{U\left(\omega_g L_s + L_m A\dfrac{d\omega_g}{dt} + L_m(\omega_g - \omega_r)B\right)^2\omega_g L_m\left(\dfrac{A}{R_s}\dfrac{dU}{dt} - \left(\dfrac{A}{R_s}\dfrac{d\omega_g}{dt} + (\omega_g - \omega_r)\dfrac{B}{R_s}\right)\dfrac{U}{\omega_g}\right)}{(R_s - \omega_g L_m(\omega_g - \omega_r)A)\left[(R_s - L_m\omega_g(\omega_g - \omega_r)A)^2 + \left(\omega_g L_s + L_m A\dfrac{d\omega_g}{dt} + L_m(\omega_g - \omega_r)B\right)^2\right]}$$

Due to $R_s \approx 0$ when $P >$ HP[12] and considering low slip, some simplifications can be done:

$$P \cong -\frac{3U L_m\left(A\dfrac{dU}{dt} - A\dfrac{d\omega_g}{dt}\dfrac{U}{\omega_g} - (\omega_g - \omega_r)B\dfrac{U}{\omega_g}\right)}{2R_s L_s}$$

$$\cong \frac{3U^2 L_m(\omega_g - \omega_r)B}{2R_s L_s\omega_g} - \frac{3U L_m A}{2R_s L_s}\dfrac{dU}{dt} + \frac{3U^2 L_m A}{2R_s L_s\omega_g}\dfrac{d\omega_g}{dt}$$

$$P \cong K_{01}U^2 + K_{02}\dfrac{U^2}{\omega_g + \Delta\omega_g} + K_1U\dfrac{dU}{dt} + K_2\dfrac{U^2}{\omega_g + \Delta\omega_g}\dfrac{d\omega_g}{dt}$$

$$Q \cong -\frac{\dfrac{3}{2}U\left(\dfrac{A}{R_s}\dfrac{dU}{dt} - \left(\dfrac{A}{R_s}\dfrac{d\omega_g}{dt} + (\omega_g - \omega_r)\dfrac{B}{R_s}\right)\dfrac{U}{\omega_g}\right)}{(\omega_g - \omega_r)A} +$$

$$+ \frac{\dfrac{3}{2}U^2}{\omega_g L_s + L_m A\dfrac{d\omega_g}{dt} + L_m(\omega_g - \omega_r)B} + \frac{\dfrac{3}{2}U\left(\dfrac{A}{R_s}\dfrac{dU}{dt} - \left(\dfrac{A}{R_s}\dfrac{d\omega_g}{dt} + (\omega_g - \omega_r)\dfrac{B}{R_s}\right)\dfrac{U}{\omega_g}\right)}{(\omega_g - \omega_r)A}$$

$$Q \cong \frac{\frac{3}{2}U^2}{\omega_g L_s + L_m A \frac{d\omega_g}{dt} + L_m(\omega_g - \omega_r)B} = \frac{\frac{3}{2}U^2}{(\omega_g + \Delta\omega_g)(L_s + L_m B) + L_m A \frac{d\omega_g}{dt} - L_m \omega_r B}$$

Acknowledgements

This work was supported by National University of Patagonia San Juan Bosco.

Author details

Roberto Daniel Fernández[1*], Pedro Eugenio Battaiotto[2] and Ricardo Julián Mantz[3]

*Address all correspondence to: dfernandez@unpata.edu.ar

1 National University of Patagonia San Juan Bosco, Argentina

2 National University of La Plata, Argentina

3 National University of La Plata and Scientific Investigation Comission of Buenos Aires State (CICpBA), Argentina

References

[1] DeMeo E, Grant W, Schuerger N. Wind Plant Integration. Costs, Status, and Issues. IEEE Power & Energy Magazine. 2005 Nov-Dec; p. 38–46.

[2] Machowski J, Bialek J, Bumby J. Power Systems Dynamics. Stability and Control. Wiley; 2008.

[3] Ummels BC, Gibescu M, Pelgrum E, Kling WL, Brand AN. Impacts of Wind Power on Thermal Generation Unit Commitment and Dispatch. IEEE Transactions on Energy Conversion. 2007 March; 22(01):44–51.

[4] Ortega-Vazquez MA, Kirschen DS. Estimating the Spinning Reserve Requirements in Systems With Significant Wind Power Generation Penetration. IEEE Transactions on Power Systems. 2009 Feb; 24 (01):144–124.

[5] Kundur P. Power Systems Stability and Control. McGraw-Hill; 1993.

[6] Sauer PW, Pai MA. Power Systems Dynamics and Stability. Prentice-Hall; 1998.

[7] Slootweg JG, Kling WL. The impact of large scale wind power generation on power system oscillations. Electric Power Systems Research. 2003;67:9–20.

[8] Fernández RD, Mantz RJ, Battaiotto PE. Potential contribution of wind farms to damp oscillations in weak grids with high wind penetration. Renewable & Sustainable EnergyReviews. 2007;doi:10.1016/j.rser.2007.01.013.

[9] Fernández RD, Mantz RJ, Battaiotto PE. Impact of Wind Farms Voltage Regulation on the Stability of the Network Frequency. International Journal of Hydrogen Energy. 2008; 33(13): 3543–3548.

[10] Fernández RD, Mantz RJ, Battaiotto PE. Wind farm non-linear control for damping electromechanical oscillations of power systems. Renewable Energy doi:101016/jrenene200801004. 2008 October;33(10):2258–2265.

[11] Zavadil R, Miller N, Muljadi E. Making Connections. IEEE Power and Energy Magazine. 2005;p. 26–37.

[12] Vas P. Sensorless Vector and Direct Torque Control. Oxford University Press; 1998.

[13] Krause K, Wasyncuk O, Sudhoff S. Analysis of Electric Machinery and Drives Systems. Wiley-Interscience IEEE Press; 2002.

[14] Ledesma P. Parques Eólicos. Doctoral Dissertation. Universidad Carlos III. Madrid. España; 2001.

[15] Mohan N. Advanced Electric Drives. Analysis, Control and Modeling using Simulink. MNPERE; 2001.

[16] Petersson A, Harnefors L, Thiringer T. Comparison between statorflux and grid-fuxoriented rotor current control of doubly-fed induction generators. In: 35th Annual IEEE Power Electronics Specialists Conference; 2004. p. 482–489. 22

[17] Thiringer T, Petersson A, Petru T. Grid disturbance response of wind turbines equipped with induction generators and Doubly-fed induction generators. IEEE Power Engineering Society General Meeting, 2003; p. 1341–1655.

[18] Xiao L, Lin L, Liu Y. Discussion on the architecture and operation mode of future power grids. Energies. 2011;(4):1025–1035.

[19] Fernández RD, Mantz RJ, Battaiotto PE. Linear and non-linear control of wind farms. Contribution to the grid stability. International Journal of Hydrogen Energy, 2010, v. 35, n. 11, p. 6019-6024.

[20] Hansen AD, Sørensen P, Iovb F, Blaabjerg F. Centralised Power Control of Wind Farm with Doubly Fed Induction Generators. Renewable Energy. 2006 June;31(7): 935–951.

[21] Gandhari M. Control Lyapunov Functions: A Control Strategy for Damping of Power Oscillations in Large Power Systems. Doctoral Dissertation. Royal Institute of Technology. Stockholm. Sweden; 2000.

[22] Freeman RA, Kokotovic PV. Robust Nonlinear Control Design. Birkhauser; 1996.

[23] Noroozian M, Ghandhari M, Andersson G, Gronquist J, Hiskens I. A Robust Control Strategy for Shunt and Series Reactive Compensators to Damp Electromechanical Oscillations. IEEE Transactions on Power Delivery. 2001 Oct;16(4):812–817.

[24] American Wind Energy Association AWEA. Electrical Guide to Utility Scale Wind Turbines; 2005. http://www.awea.org. March 27nd, 2007. AWEA Grid Code White Paper.

[25] Ortega R, García Canseco E. Interconection and Damping Assignment Passivity-Based Control: Towards a Constructive Procedure - Part I. 43rd IEEE Conference on Decision and Control. 2004; ThB07.1.

[26] Ortega R, García Canseco E. Interconection and Damping Assignment Passivity-Based Control: Towards a Constructive Procedure - Part II. 43rd IEEE Conference on Decision and Control. 2004; ThB07.2.

[27] Ortega R, Loría A, Nicklasson PJ, Sira-Ramírez H. Passibity-based Control of Euler-Lagrange Systems. Springer-Verlag; 1998.

[28] Galaz M, Ortega R, Bazanella AS, Stankovic AM. An energy-shaping approach to the design of excitation control of synchronous generators. Automatica. 2003; 39(1):111–119.

[29] Ortega R, Galaz M, Astolfi A, Sun Y, Shen T. Transient Stabilization of Multimachine Power Systems with Nontrivial Transfer Conductances. IEEE Transactions on Automatic Control. 2005 Jan; 50(1):60–75.

[30] Fernández RD, Mantz RJ, Battaiotto PE. Passivity based wind farm control for transient stabilization of an electrical network. In: World Wind Energy Conference & Exhibition -WWEC 2007; 2007.

[31] Fernández RD, Mantz RJ, Battaiotto PE. Wind farm control for stabilisation of electrical networks based on passivity. International Journal of Control. 2010;83(1):105–114.

Chapter 6

An Integrated Power Flow Solution of Flexible AC Transmission Systems Containing Wind Energy Conversion Systems

E. Barrios-Martinez, L.M. Castro,
C.R. Fuerte-Esquivel and C. Angeles-Camacho

Additional information is available at the end of the chapter

1. Introduction

Even though the use of wind generators for converting wind energy into electricity is beneficial from the environmental standpoint, their consideration in the active power dispatch makes the already complex task of achieving power system controllability even more demanding. Consequently, the quantification of the effects that large-scale integration of wind generation will cause on the network is a very important matter that requires special attention when planning and operating an electrical power system. Arguably, power flow analysis is the most popular computational calculation performed in a power system's planning and operation, and this study has been selected to quantify the electrical response of wind generators when they are integrated in Flexible AC Transmission Systems.

Mathematical models of several types of wind generators have been developed in which their active and reactive power outputs are obtained based on the steady-state equivalent representation of the induction machine. The power injection method is then used to include these models into the power flow formulation, which is solved by using a sequential approach to obtain an operating point of the power system. In this approach, only the network's state variables are calculated through a conventional power flow algorithm, while a subproblem is formulated for updating the state variables of wind generators as well as their power injections at the end of each power flow's iteration.

Instead of using the power injection concept, another way of representing a wind generator is by means of an equivalent variable impedance expressed in terms of the slip of the generator and its rotor and stator winding parameters. This impedance is included in the system's admittance matrix, and the network nodal voltages are computed through the power flow analysis. Based on these voltages, the air-gap power of the wind generator is calculated and used to iteratively compute the value of the slip of the induction generator that produces the match between the air-gap power and the mechanical power extracted from the wind.

In general terms, all the methods discussed above share the characteristic of using a sequential approach to calculate the state variables of the wind generators, and none of them considers the integration of FACTS controllers in the network where the WECSs are embedded.

A fundamentally different approach for the modelling of WECS, within the context of the power flow problem, is a method that simultaneously combines the state variables associated with the wind generators, the FACTS controllers and the transmission network in a single frame-of-reference for a unified iterative solution through a Newton-Raphson (NR) technique. From the convergence standpoint, the unified method is superior to the sequential one because the interaction between the network, FACTS controllers and wind generators is better represented during the iterative solution. Furthermore, it arrives at the solution with a quadratic convergence regardless of the network size. Hence, the key contribution of this work is to provide a comprehensive and general approach for the analysis of power flows in Flexible AC Transmission Systems containing wind generators in a unified single-frame of reference.

2. Power flow including FACTS controllers and wind generators

The unified approach suggested in [1] is extended to compute the power flow solution of a power system containing FACTS controllers and WECS furthermore, the approach is represented by a single set of nonlinear power flow mismatch equations $f(X_{AC}, X_F, X_{WG})=0$, where X_{AC} is a vector of all nodal voltage magnitudes and angles, X_F stands for the state variables of the FACTS controllers and X_{WG} is a vector of all state variables associated with the wind generators. The linearised power flow mismatch equations corresponding to the wind farms are then combined with those associated with the FACTS controllers and the rest of the network, as given by (1), which are solved iteratively by the NR method:

$$
\begin{bmatrix} \Delta P \\ \Delta Q \\ \Delta R_F \\ \Delta R_{WG} \end{bmatrix}^j =
\begin{bmatrix}
\dfrac{\partial P}{\partial \theta} & V\dfrac{\partial P}{\partial V} & \dfrac{\partial P}{\partial X_F} & \dfrac{\partial P}{\partial X_{WG}} \\[2ex]
\dfrac{\partial Q}{\partial \theta} & V\dfrac{\partial Q}{\partial V} & \dfrac{\partial Q}{\partial X_F} & \dfrac{\partial Q}{\partial X_{WG}} \\[2ex]
\dfrac{\partial R_F}{\partial \theta} & V\dfrac{\partial R_F}{\partial V} & \dfrac{\partial R_F}{\partial X_F} & \dfrac{\partial R_F}{\partial X_{WG}} \\[2ex]
\dfrac{\partial R_{WG}}{\partial \theta} & V\dfrac{\partial R_{WG}}{\partial V} & \dfrac{\partial R_{WG}}{\partial X_F} & \dfrac{\partial R_{WG}}{\partial X_{WG}}
\end{bmatrix}^j
\begin{bmatrix} \Delta\theta \\ \dfrac{\Delta V}{V} \\ \Delta X_F \\ \Delta X_{WG} \end{bmatrix}^j
\tag{1}
$$

where ΔR_F and ΔR_{WG} represent the mismatch equations of the FACTS controllers and wind generators, respectively. The NR method starts from an initial guess for all the state variables and updates the solution at each iteration i until a predefined tolerance is fulfilled. In this unified solution, all the state variables are adjusted simultaneously in order to compute the steady-state operating condition of the power system. Hence, this method has strong convergence characteristics.

3. Modelling of FACTS devices

Among all FACTS devices used to improve the steady-state performance of power systems [1], the Static Var Compensator (SVC) and Thyristor-Controlled Series Capacitor (TCSC) are the controllers considered.

3.1. Static VAR compensator

An integrated SVC and step-down transformer model is obtained by combining the admittances of both components $Y_{T\text{-}SVC} = Y_T + Y_{SVC}$ as proposed in [1]. The linearised power flow equations are given by (2) considering the firing angle α_{SVC} of the SVC as the state variable within the NR method

$$\begin{bmatrix} \Delta P_k \\ \Delta Q_k \end{bmatrix} = \begin{bmatrix} 0 & \dfrac{\partial G_{T\text{-}SVC}}{\partial \alpha_{SVC}} V_k^2 \\ 0 & -\dfrac{\partial B_{T\text{-}SVC}}{\partial \alpha_{SVC}} V_k^2 \end{bmatrix} \begin{bmatrix} \Delta \theta_k \\ \Delta \alpha_{SVC} \end{bmatrix} \tag{2}$$

where $G_{T\text{-}SVC}$ and $B_{T\text{-}SVC}$ are functions dependent on α_{SVC}:

$$G_{T\text{-}SVC} = \frac{R_T}{R_T^2 + X_{Eq}^2},$$

$$B_{T\text{-}SVC} = -\frac{X_{Eq}}{R_T^2 + X_{Eq}^2},$$

$$X_{Eq} = X_T + X_{SVC},$$

$$X_{SVC} = \frac{X_C X_{TCR}}{X_C - X_{TCR}},$$

and

$$X_{TCR} = \frac{\pi X_L}{2(\pi - \alpha_{SVC}) + \sin(2\alpha_{SVC})}.$$

3.2. Thyristor-controlled series compensator

The TCSC firing angle power flow model is represented as an equivalent series reactance which is associated with the firing angle α_{TCSC}. This reactance can be expressed as [1] (see Appendix):

$$
\begin{aligned}
X_{TCSC} = -X_C + C_1 \left\{ 2(\pi - \alpha) + \sin\left[2(\pi - \alpha) \right] \right\} \\
- C_2 \cos^2(\pi - \alpha) \left\{ \varpi \tan\left[\varpi(\pi - \alpha) \right] - \tan(\pi - \alpha) \right\}
\end{aligned}
\tag{3}
$$

Assuming the TCSC controls the active power flowing from bus k to bus m to a specified value of P_{km}^{sp}, the set of linearised power flow equations is

$$
\begin{bmatrix} \Delta P_k \\ \Delta Q_k \\ \Delta P_m \\ \Delta Q_m \\ \Delta P_{km} \end{bmatrix}^i
=
\begin{bmatrix}
\frac{\partial P_k}{\partial \theta_k} & \frac{\partial P_k}{\partial V_k}V_k & \frac{\partial P_k}{\partial \theta_m} & \frac{\partial P_k}{\partial V_m}V_m & \frac{\partial P_k}{\partial \alpha_{TCSC}} \\
\frac{\partial Q_k}{\partial \theta_k} & \frac{\partial Q_k}{\partial V_k}V_k & \frac{\partial Q_k}{\partial \theta_m} & \frac{\partial Q_k}{\partial V_m}V_m & \frac{\partial Q_k}{\partial \alpha_{TCSC}} \\
\frac{\partial P_m}{\partial \theta_k} & \frac{\partial P_m}{\partial V_k}V_k & \frac{\partial P_m}{\partial \theta_m} & \frac{\partial P_m}{\partial V_m}V_m & \frac{\partial P_m}{\partial \alpha_{TCSC}} \\
\frac{\partial Q_m}{\partial \theta_k} & \frac{\partial Q_m}{\partial V_k}V_k & \frac{\partial Q_m}{\partial \theta_m} & \frac{\partial Q_m}{\partial V_m}V_m & \frac{\partial Q_m}{\partial \alpha_{TCSC}} \\
\frac{\partial P_{km}^\alpha}{\partial \theta_k} & \frac{\partial P_{km}^\alpha}{\partial V_k}V_k & \frac{\partial P_{km}^\alpha}{\partial \theta_m} & \frac{\partial P_{km}^\alpha}{\partial V_m}V_m & \frac{\partial P_{km}^\alpha}{\partial \alpha_{TCSC}}
\end{bmatrix}^i
\begin{bmatrix} \Delta \theta_k \\ \frac{\Delta V_k}{V_k} \\ \Delta \theta_m \\ \frac{\Delta V_m}{V_m} \\ \Delta \alpha_{TCSC} \end{bmatrix}^i
\tag{4}
$$

where the power flow mismatch for the TCSC module is defined as $\Delta P_{km} = P_{km}^{sp} - P_{km}^{\alpha}$, $P_{km}^{\alpha} = V_k V_m B_{km} \sin(\theta_k - \theta_m)$ and $B_{km} = 1/X_{TCSC}$.

4. Modelling of wind generators

Wind generators are categorized according to how they operate when they are connected to the grid. The Fixed-Speed Wind Generators (FSWG) are thus named so because their speed is mainly set according to the system's frequency [2]. In this category are the Stall-Regulated Fixed-Speed Wind Generators (SR-FSWG) and the Pitch-Regulated Fixed-Speed Wind Generators (PR-FSWG). A variant of the previous models is the semi-variable speed wind generator (SSWG), which uses a wound-rotor induction generator with an external resistor added to the rotor circuit in order to achieve a power regulation when wind speeds are above the rated one [3]. Also, variable-speed wind generators are being employed worldwide with the doubly-fed induction generator being the most used. However, another emergent topology that is being widely accepted is the wind generator based on a Permanent Magnet Synchronous Generator (PMSG) with a full-scale converter in which the gearbox can be omitted [4].

Mathematical modelling of FSWG, SSWG and PMSG-based wind generators for power flow studies is addressed below. In reference [5], the models of fixed- and semi-variable speed wind generators are suitably derived for power flow analysis and can be readily integrated in the formulation presented herein. For this reason, only a brief description of these models is given next.

4.1. Fixed-speed wind generators

This generator is directly connected to the network through a step-up transformer, and its final operating point depends upon the electrical frequency as well as the nodal voltage at the generator's terminals. The generated reactive and active powers are determined by Equations (5) and (6), respectively, and the stator and rotor currents of the induction genera-tor can be expressed according to Equations (7) and (8) [5] :

$$Q_g(V,s) = -V^2 \left[\frac{A + B s^2}{(C - D s)^2 + (E + F s)^2} \right] \tag{5}$$

$$P_g(V,s) = -V^2 \left[\frac{K + H s + L s^2}{(C - D s)^2 + (E + F s)^2} \right] \tag{6}$$

$$I_1^2(V,s) = V^2 \left\{ \frac{\left(K + H s + L s^2\right)^2 + \left(A + B s^2\right)^2}{\left[(C - D s)^2 + (E + F s)^2\right]^2} \right\} \tag{7}$$

$$I_2^2(V,s) = V^2 \left\{ \frac{\left(M s + N s^2\right)^2 + \left(T s - W s^2\right)^2}{\left[(C - D s)^2 + (E + F s)^2\right]^2} \right\} \tag{8}$$

where s is the machine's slip, V is the terminal voltage, and the constants from A to W are as given in the Appendix. Furthermore, the power converted from mechanical to electrical form (P_{conv}) can be computed by using Equation (9), where R_2 represents the rotor resistance

$$P_{conv} = -I_2^2 R_2 \left(\frac{1-s}{s} \right) \tag{9}$$

4.1.1. Stall-regulated fixed-speed wind generator

The mechanical power P_m [W] extracted from the wind by this generator is given by [6]

$$P_m = 0.5 \cdot \rho \cdot c_1 \left(\frac{c_2}{\lambda_i} - c_3\beta - c_4\beta^{c_5} - c_6 \right) \cdot e^{\frac{-c_7}{\lambda_i}} \cdot A \cdot V_w^3 \tag{10}$$

where

$$\lambda_i = \left[\left(\frac{1}{\lambda + c_8\beta} \right) - \left(\frac{c_9}{\beta^3 + 1} \right) \right]^{-1}$$

and

$$\lambda = \frac{R \cdot n_{gb} \cdot \omega_S (1-s)}{V_w}$$

- ρ is the air density [kg/m³],

- A is the swept area of the blades [m²],

- V_w is the wind speed [m/s],

- R is the radius of the rotor [m],

- n_{gb} is the gearbox ratio,

- ω_s is the angular synchronous speed [rad/s],

- β is the pitch angle [degrees],

- ω_T is the angular speed of the turbine [rad/s],

- and the constants c_1 to c_9 are the parameters of the wind turbine's design.

Thus, assuming that the SR-FSWG is connected at bus k, the power mismatches equations are (11)-(13), and the set of linearised equations that has to be assembled and combined with the Jacobian matrix and the power mismatch vector of the entire network is shown in Equation (14) [5]:

$$\Delta P_k = P_g(V,s) - P_{Lk} - P_k^{cal} = 0 \tag{11}$$

$$\Delta Q_k = Q_g(V,s) - Q_{Lk} - Q_k^{cal} = 0 \tag{12}$$

$$\Delta P_{WT1,k} = -\{P_m - P_{conv}\} = -\left\{ P_m + I_2^2 R_2 \left(\frac{1-s}{s} \right) \right\} = 0 \tag{13}$$

$$
\begin{bmatrix} \Delta P_k \\ \Delta Q_k \\ \Delta P_{WT1,k} \end{bmatrix}^j = \begin{bmatrix} \dfrac{\partial P_k^{cal}}{\partial \theta_k} & \left(\dfrac{\partial P_k^{cal}}{\partial V_k} - \dfrac{\partial P_g}{\partial V_k} \right) V_k & \dfrac{\partial P_g}{\partial s} \\[2ex] \dfrac{\partial Q_k^{cal}}{\partial \theta_k} & \left(\dfrac{\partial Q_k^{cal}}{\partial V_k} - \dfrac{\partial Q_g}{\partial V_k} \right) V_k & \dfrac{\partial Q_g}{\partial s} \\[2ex] 0 & \dfrac{\partial P_{WT1,k}}{\partial V_k} V_k & \dfrac{\partial P_{WT1,k}}{\partial s} \end{bmatrix}^j \begin{bmatrix} \Delta \theta_k \\ \dfrac{\Delta V_k}{V_k} \\ \Delta s \end{bmatrix}^j
\tag{14}
$$

where $P_g(V,s)$ and $Q_g(V,s)$ are given by (6) and (5), respectively, P_{LK} and Q_{LK} represent the active and reactive powers drawn by the load at bus k, respectively, and P_k^{cal} and Q_k^{cal} are active and reactive power injections given by

$$
P_k^{cal} = V_k^2 G_{kk} + V_k \sum_{m \in k} V_m \left[G_{km} \cos(\theta_k - \theta_m) + B_{km} \sin(\theta_k - \theta_m) \right]
\tag{15}
$$

$$
Q_k^{cal} = -V_k^2 B_{kk} + V_k \sum_{m \in k} V_m \left[G_{km} \sin(\theta_k - \theta_m) - B_{km} \cos(\theta_k - \theta_m) \right]
\tag{16}
$$

4.1.2. Pitch-regulated fixed-speed wind generator

Since this wind generator has a blade pitch angle mechanism which actuates to limit the power extracted from the wind [7], the generated active power $P_{g,pr}$ can be obtained from its power curve and is considered constant at a value $P_{g,pr}^{sp}$ through the iterative process; however, the reactive power $Q_{g,pr}=Q_g$ needs to be calculated [5]. Therefore, the internal power equilibrium point in the wind generator has to be computed by Equation (17):

$$
P_{m,pr} = P_{g,pr}^{sp} + P_{losses,s} + P_{losses,r} = P_{g,pr}^{spe} + 3I_1^2 R_1 + 3I_2^2 R_2
\tag{17}
$$

where $P_{losses,s}$ and $P_{losses,r}$ are the three-phase stator and rotor power losses, respectively, and the core losses in the induction machine are neglected. As mentioned previously, the set of linearised power flow mismatch equations regarding the PR-FSWG is given by Equations (18)-(21) when the generator is connected at bus k:

$$
\Delta P_k = P_{g,pr} - P_{Lk} - P_k^{cal} = 0
\tag{18}
$$

$$
\Delta Q_k = Q_g(V,s) - Q_{Lk} - Q_k^{cal} = 0
\tag{19}
$$

$$
\Delta P_{WT2,k} = -P_{g,pr}^{sp} - \left\{ \left(3I_1^2 R_1 + 3I_2^2 R_2 \right) + I_2^2 R_2 \left(\frac{1-s}{s} \right) \right\} = 0
\tag{20}
$$

$$\begin{bmatrix} \Delta P_k \\ \Delta Q_k \\ \Delta P_{WT2,k} \end{bmatrix}^j = \begin{bmatrix} \dfrac{\partial P_k^{cal}}{\partial \theta_k} & \dfrac{\partial P_k^{cal}}{\partial V_k}V_k & 0 \\[3mm] \dfrac{\partial Q_k^{cal}}{\partial \theta_k} & \left(\dfrac{\partial Q_k^{cal}}{\partial V_k} - \dfrac{\partial Q_g}{\partial V_k}\right)V_k & \dfrac{\partial Q_g}{\partial s} \\[3mm] 0 & \dfrac{\partial P_{WT2,k}}{\partial V_k}V_k & \dfrac{\partial P_{WT2,k}}{\partial s} \end{bmatrix}^j \begin{bmatrix} \Delta \theta_k \\ \dfrac{\Delta V_k}{V_k} \\ \Delta s \end{bmatrix}^j \tag{21}$$

4.2. Semi-variable speed wind generator

In this type of generator, the slip of the induction machine cannot be regarded as the state variable because of the external resistance R_{ext} added in the rotor circuit [5]. Hence, a total rotor circuit resistance, $R_x = (R_2 + R_{ext})/s$, is considered as a single-state variable associated with the rotor circuit, which is adjusted to satisfy the power mismatch equations during the NR power flow calculation [5,8]. Hence, the reactive and active powers, the stator and rotor currents as well as the power converted from mechanical to electrical form will be dependent functions on R_x and can be expressed as

$$Q_{g,ss}(V,R_x) = -V^2\left[\frac{A'R_x^2 + B}{(C'R_x - D)^2 + (E'R_x + F)^2}\right] \tag{22}$$

$$P_{g,ss}(V,R_x) = -V^2\left[\frac{K'R_x^2 + H'R_x + L}{(C'R_x - D)^2 + (E'R_x + F)^2}\right] \tag{23}$$

$$I_{1,ss}^2(V,R_x) = V^2\left\{\frac{\left(K'R_x^2 + H'R_x + L\right)^2 + \left(A'R_x^2 + B\right)^2}{\left[(C'R_x - D)^2 + (E'R_x + F)^2\right]^2}\right\} \tag{24}$$

$$I_{2,ss}^2(V,R_x) = V^2\left\{\frac{(M'R_x + N)^2 + (T'R_x - W)^2}{\left[(C'R_x - D)^2 + (E'R_x + F)^2\right]^2}\right\} \tag{25}$$

$$P_{conv} = -I_{2,ss}^2 R_2\left(\frac{1-s}{s}\right) \approx -I_{2,ss}^2 R_2\left(\frac{1}{s}\right) = -I_{2,ss}^2 R_x \tag{26}$$

where the constants A', C', E', H', K', M' and T' are given in the Appendix. Since the generated active power $P_{g,ss}$ is set to a fixed value $P_{g,ss}^{sp}$ obtained from the wind generator power curve, and assuming no core losses, the mechanical power of the induction generator $P_{m,ss}$ can be estimated as follows:

$$P_{m,ss} = P_{g,ss}^{sp} + P_{losses,s} + P_{losses,r} = P_{g,ss}^{sp} + 3I_{1,ss}^2 R_1 + 3I_{2,ss}^2 R_2 \qquad (27)$$

Therefore, when the SSWG is connected at bus k, its set of mismatch power flow equations is

$$\Delta P_k = P_{g,ss}^{sp} - P_{Lk} - P_k^{cal} = 0 \qquad (28)$$

$$\Delta Q_k = Q_{g,ss}(V, R_x) - Q_{Lk} - Q_k^{cal} = 0 \qquad (29)$$

$$\Delta P_{WT3,k} = -P_{g,ss}^{sp} - \left\{ 3I_{1,ss}^2 R_1 + I_{2,ss}^2 R_x \right\} = 0 \qquad (30)$$

$$
\begin{bmatrix} \Delta P_k \\ \Delta Q_k \\ \Delta P_{WT3,k} \end{bmatrix}^j =
\begin{bmatrix}
\dfrac{\partial P_k^{cal}}{\partial \theta_k} & \dfrac{\partial P_k^{cal}}{\partial V_k} V_k & 0 \\
\dfrac{\partial Q_k^{cal}}{\partial \theta_k} & \left(\dfrac{\partial Q_k^{cal}}{\partial V_k} - \dfrac{\partial Q_{g,ss}}{\partial V_k} \right) V_k & \dfrac{\partial Q_{g,ss}}{\partial R_x} V_k \\
0 & \dfrac{\partial P_{WT3,k}}{\partial V_k} V_k & \dfrac{\partial P_{WT3,k}}{\partial R_x}
\end{bmatrix}^j
\begin{bmatrix} \Delta \theta_k \\ \dfrac{\Delta V_k}{V_k} \\ \Delta R_x \end{bmatrix}^j \qquad (31)
$$

where the internal energy balance in the induction machine $\Delta P_{WT3} = -(P_{m,ss} - P_{conv})$ is derived by using Equations (26)-(27).

4.3. PMSG-based wind generator

This type of wind generator possesses a PMSG and a full-rated converter to connect the generator to the network, resulting in complete speed and reactive power control [9]. Hence, all the generated power is supplied to the power system through a machine-side converter and grid-side converter. The schematic diagram of this topology is shown in figure 1(a). Reactive power support is one of the characteristics that make this machine attractive for wind power production. In this case, the inclusion of the explicit representation of the wind generator step-up transformer is considered, which allows for direct voltage magnitude control at the high-voltage side of the transformer. The proposed model for power flow studies is shown in the figure 1(b) in which $P_{g,pmsg}$ represents the output power set by the wind generator power curve for a given wind speed, V_{msc} and V_{gsc} are the voltage at the machine-side converter and grid-side converter terminal, respectively, and Z_{st} is the step-up transformer impedance.

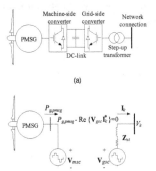

Figure 1. PMSG-based wind generator: (a) schematic diagram, (b) proposed model for power flow studies.

The power flow equations for the PMSG-based wind generator are derived assuming the following voltage at the grid-side converter terminal: $V_{gsc} = V_{gsc} \, e^{\,j\delta_{gsc}}$. Based on Fig. 1(b), the active and reactive powers flowing from the grid-side converter terminal to the k-th bus are

$$P_{gsc-k} = V_{gsc}^2 G_{st} + V_{gsc} V_k \left[G_{st} \cos\left(\delta_{gsc} - \theta_k \right) + B_{st} \sin\left(\delta_{gsc} - \theta_k \right) \right] \tag{32}$$

$$Q_{gsc-k} = -V_{gsc}^2 B_{st} + V_{gsc} V_k \left[G_{st} \sin\left(\delta_{gsc} - \theta_k \right) - B_{st} \cos\left(\delta_{gsc} - \theta_k \right) \right] \tag{33}$$

For the active and reactive powers at bus k, the subscripts gsc and k are exchanged in Equations (32) and (33). Therefore, the NR-based power flow formulation is given by

$$\Delta P_k = -P_{k-gsc} - P_{Lk} - P_k^{cal} = 0 \tag{34}$$

$$\Delta Q_k = -Q_{k-gsc} - Q_{Lk} - Q_k^{cal} = 0 \tag{35}$$

$$\Delta P_{WT4,k} = P_{g,pmsg} - P_{gsc-k} = 0 \tag{36}$$

$$
\begin{bmatrix} \Delta P_k \\ \Delta Q_k \\ \Delta P_{WT4,k} \end{bmatrix}^j =
\begin{bmatrix}
\dfrac{\partial P_k^{cal}}{\partial \theta_k} & \dfrac{\partial P_k^{cal}}{\partial V_{gsc}} V_{gsc} & \dfrac{\partial P_{k-gsc}}{\partial \delta_{gsc}} \\[2ex]
\dfrac{\partial Q_k^{cal}}{\partial \theta_k} & \dfrac{\partial Q_k^{cal}}{\partial V_{gsc}} V_{gsc} & \dfrac{\partial Q_{k-gsc}}{\partial \delta_{gsc}} \\[2ex]
\dfrac{\partial P_{WT4,k}}{\partial \theta_k} & \dfrac{\partial P_{WT4,k}}{\partial V_{gsc}} V_{gsc} & \dfrac{\partial P_{WT4,k}}{\partial \delta_{gsc}}
\end{bmatrix}^j
\begin{bmatrix} \Delta \theta_k \\ \dfrac{\Delta V_{gsc}}{V_{gsc}} \\ \Delta \delta_{gsc} \end{bmatrix}^j \tag{37}
$$

Note that Equation (36) represents the power constraint in the AC/DC/AC converter in which active power losses are neglected.

5. Case studies with wind farms and FACTS devices

This section shows how the proposed approach performs when considering a power system having FACTS devices and wind generators.

5.1. Five-bus test system with FSWGs, SSWGs and a SVC

The typical five-bus test system is used to provide an example with the inclusion of a wind farm consisting of ten SR-FSWGs, ten PR-FSWGs and ten SSWGs operating at a wind speed of 16 m/s with which all wind generators are injecting their maximum power. Additionally, a SVC is placed at bus five in order maintain its terminal voltage magnitude at 1 pu. The conventional generators are set to control voltage magnitudes at 1 pu. Parameters of the wind farm are given in the Appendix.

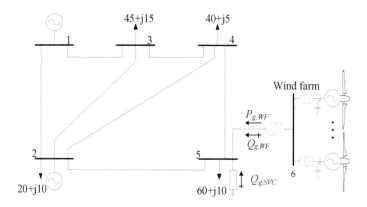

Figure 2. Modified five-bus test system used to incorporate a SVC and a wind farm composed of several wind generator models.

In order to show the effect of wind farms and a SVC in the operation of the power system, the following three scenarios are presented: (a) the base case where the wind farm and SVC are not considered, (b) the case where only the wind farm is running and (c) the case where the SVC is connected at the bus at which the wind farm is connected in order to provide voltage support. The results regarding each case are reported in Table I.

Results	Scenario		
	(a)	(b)	(c)
V_1	1.000	1.000	1.000
V_2	1.000	1.000	1.000
V_3	0.971	0.971	0.977
V_4	0.971	0.971	0.979
V_5	0.967	0.966	1.000
V_6	---	0.945	0.981
$P_{g,WF}$	---	25.349	25.341
$Q_{g,WF}$	---	-10.802	-9.857
$Q_{g,SVC}$	---	---	36.897

Table 1. Power flow simulation results with wind farm and SVC.

The simulated wind farm is a reactive power consumer since it lacks a reactive power control as seen from Table 1. When no SVC is considered, its reactive power absorption exceeds 10 MVAr, resulting in a low-voltage magnitude at node six. On the other hand, when the SVC is set in operation, not only the low-voltage side of the wind farm transformer is boosted, but also the system voltage profile.

Since all wind generators are operating at the same wind speed, clearly the state variables calculated for each group of wind generators corresponding to each model will result in the same value. Furthermore, if the system voltage profile changes, as occurs with the inclusion of the SVC, another operating point is found at each wind generator as shown in Table 2.

Scenario	SR-FSWG	PR-FSWG	SSWG	SVC
s	s	R_x	a_{SVC} (deg)	
(b)	-0.00506	-0.00679	-67.69437	---
(c)	-0.00467	-0.00613	-75.16952	136.35024

Table 2. Computed wind generators and SVC state variables.

5.2. Five-bus test system with PMSG-based wind generators and a TCSC

In this case, a PMSG-based wind farm is located at bus 4 with 30 wind generators operating at a rated wind speed, i.e. 15 m/s. Be aware that each PMSG-based wind generator provides reactive power support by controlling its terminal voltage magnitude. Also, a TCSC is placed for controlling the active power flowing through the transmission line connected between nodes 4 and 5 at P_{4-5}= 20 MW, as shown in figure 2. The next scenarios are analyzed: (a) the base case where the wind farm and TCSC are not considered, (b) the power system including only the PMSG-based wind farm and (c) the network having simultaneously the TCSC and the wind farm. The results are reported in Table 3.

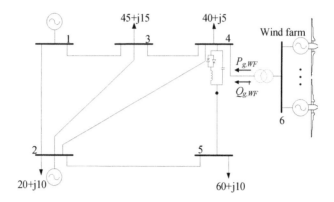

Figure 3. Modified five-bus test system used to incorporate a TCSC and a PMSG-based wind farm.

Results		Scenario	
(a)	(b)	(c)	
V_1	1.000	1.000	1.000
V_2	1.000	1.000	1.000
V_3	0.971	0.984	0.984
V_4	0.971	0.987	0.988
V_5	0.967	0.973	0.971
V_6	---	1.000	1.000
$P_{g,WF}$	---	60.000	60.000
$Q_{g,WF}$	---	2.566	2.218
P_{1-2}	89.683	46.792	45.883
P_{1-3}	40.639	20.252	21.206
P_{2-3}	24.821	11.565	13.114
P_{2-4}	28.076	9.279	11.253
P_{2-5}	55.032	45.465	41.053
P_{4-3}	-18.686	13.628	11.171
P_{4-5}	6.258	15.590	20.000

Table 3. Power flow simulation results with wind farm and TCSC.

The system voltage profile is improved when the PMSG-based wind farm is integrated to the grid as seen from Table 3. This is mainly due to two reasons; there is a redistribution of power flows in the network, e.g. the load connected at node four is being supplied locally by the wind farm. On the other hand, the PMSG-based wind farm is providing voltage support, resulting in a voltage magnitude of 1 pu at the low-voltage side of the wind farm transformer even when the TCSC is set in operation to increase the power transfer to 20 MW in the line connecting nodes 4 and 5. The wind generators and TCSC state variables estimated by the NR algorithm for each scenario are the following: scenario (b) $V_{gsc} = 1.017$ pu, $\delta_{gsc} = 9.071°$, and scenario (c) $V_{gsc} = 1.016$ pu, $\delta_{gsc} = 8.862°$, $\alpha_{TCSC} = 143.499°$.

6. Discussion

This chapter has put forward the NR-based power flow algorithm which is capable of computing the steady-state operating point of electric networks containing WECS and FACTS devices. The solution problem is formulated in a single-frame of reference, resulting in an efficient iterative solution. Additionally, a PMSG-based wind generator model for power flow studies is presented, which allows for direct voltage magnitude control at the high-voltage side of the wind generator transformer. Numerical examples have shown that FACTS controllers are a practical alternative to integrate WECS into power systems without degrading their operative performance.

Appendix

- TCSC parameters:

$$C_1 = \frac{X_C + X_{LC}}{\pi}, \quad C_2 = \frac{4X_{LC}^2}{\pi X_L}, \quad X_{LC} = \frac{X_C X_L}{X_C - X_L}, \quad \omega = \left(\frac{X_C}{X_L}\right)^{1/2}.$$

- FSWG parameters:

$$A = R_2^2(X_1 + X_m), \quad B = (X_2 + X_m)[X_2 X_m + X_1(X_2 + X_m)], \quad C = R_1 R_2, \quad D = X_2 X_m + X_1(X_2 + X_m),$$
$$E = R_2(X_1 + X_m), \quad F = R_1(X_2 + X_m), \quad H = R_2 X_m^2, \quad K = R_1 R_2^2, \quad L = R_1(X_2 + X_m)^2,$$
$$M = X_m R_2(X_1 + X_m), \quad N = X_m R_1(X_2 + X_m), \quad T = R_1 R_2 X_m, \quad W = X_m[X_2 X_m + X_1(X_2 + X_m)].$$

- SSWG parameters:

$$A' = (X_1 + X_m), \quad C' = R_1, \quad E' = (X_1 + X_m), \quad H' = X_m^2, \quad K' = R_1, \quad M' = X_m(X_1 + X_m), \quad T' = R_1 X_m.$$

The data for each wind farm is (on a base power of 100 MVA): wind farm step-up transformer impedance is 0.2 pu, and the impedance of each wind generator transformer is 4.1667 pu. Also, the data of each wind generator model are given in Table 4

	SR-FSWG	PR-FSWG	SSWG
Z_1	0.0027 + j0.025	0.0 + j0.09985	0.00269 + j0.072605
Z_2	0.0022 + j0.046	0.00373 + j0.10906	0.002199 + j0.04599
Z_m	j1.38	j3.54708	j1.37997
Vnom	690	690	690
Pnom	900	600	1000

Table 4. Wind generator data.

where Z_1 is the stator impedance [Ω], Z_2 is the rotor impedance [Ω], Z_m is the magnetizing impedance [Ω], V_{nom} is the rated voltage [V] and P_{nom} is the rated power of the wind generator [kW].

The coefficients of Equation (10) are as follows: $c_1 = 0.5$; $c_2 = 116$; $c_3 = 0.4$; $c_4 = 0.0$; $c_5 = 0$; $c_6 = 5$; $c_7 = 21$; $c_8 = 0.08$; and $c_9 = 0.035$; $\beta = 0$. Also, for the PMSG-based wind generator, its rated voltage is 690 V, and its rated power is 2000 kW.

Acknowledgements

The authors gratefully acknowledge the financial support granted to MSc. Luis M. Castro by the Consejo Nacional de Ciencia y Tecnología (CONACYT) México, and the University of Michoacán (U.M.S.N.H) for allowing him to undertake PhD studies. The authors gratefully acknowledge the financial support granted to Fuerte-Esquivel CR and Angeles-Camacho C by the FI and the II at the UNAM under the research project 2102.

Author details

E. Barrios-Martinez[1], L.M. Castro[2], C.R. Fuerte-Esquivel[1,2] and C. Angeles-Camacho[1]

1 Instituto de Ingeniería, Universidad Nacional Autónoma de México, UNAM, México

2 Universidad Michoacana de San Nicolás de Hidalgo, UMSNH, Michoacán, México

References

[1] Acha E, Fuerte-Esquivel CR, Ambriz-Perez H, Angeles-Camacho C. FACTS : Modelling and Simulation in Power Networks, Chichester: John Wiley & Sons; 2004.

[2] Hwang PI, Ahn SJ, Moon SI. Modeling of the Fixed Speed Wind Turbine Generator System for DTS. IEEE PES General Meeting, 2008: 1-7.

[3] Burnham DJ, Santoso S, Muljadi E. Variable Rotor-Resistance Control of Wind Turbine Generators. IEEE PES General Meeting, 2009: 1-6.

[4] Senjyu T, Yona A, Funabashi T. Operation Strategies for Stability of Gearless Wind Power Generation Systems. IEEE PES General Meeting, 2008: 1-7.

[5] Castro LM, Fuerte-Esquivel CR, Tovar-Hernández JH. A Unified Approach for the Solution of Power Flows in Electric Power Systems Including Wind Farms. Electric Power Systems Research, in press.

[6] Ackerman T. Wind Power in Power Systems, 1st ed. Chichester: John Wiley & Sons; 2005.

[7] Bianchi FD, De Battista H, Mantz RJ. Wind Turbine Control Systems - Principles, Modelling and Gain Scheduling Design, 1st ed. London: Springer-Verlan; 2006.

[8] Divya KC, Rao PSN. Models for wind turbine generating systems and their application in load flow studies. Electric Power Systems Research 2006: 76:844–856.

[9] Hong-Woo K, Sung-Soo K, Hee-Sang K. Modeling and control of PMSG-based variable-speed wind turbine. Electric Power Systems Research 2010:80:46-52.

Chapter 7

Modeling Issues of Grid-Integrated Wind Farms for Power System Stability Studies

Tamer A. Kawady and Ahmed M. Nahhas

Additional information is available at the end of the chapter

1. Introduction

Owing to the rapid increase of the global population, uneven distribution of resources and the non-renewable nature of fossil fuels, the importance of renewable energy resources is obvious. Further, growing environmental concerns and attempts to reduce dependency on fossil fuel resources are bringing renewable energy resources to the mainstream of the electric power sector. Among the various renewable resources, wind power is assumed to have the most favourable, technical and economical prospects. According to the new energy policies regarding the share of renewable energy by 2020, different countries all over the world have their own targets. Examples of these countries with their penetration levels included Denmark (52%), Sweden (50%), Spain (41%) Germany (39%) and Finland (38%). Accordingly, realizing an accurate representation of such power plants and their remarkable issues regarding their integration to the power system is highly demanded.

According to the annual report of the Global Wind Energy Council (GWEC), over 40 GW of new wind power generation capacity came on line worldwide in 2011 attracting more than $68 billions [1]. This brings the total global wind power capacity to over 238 GW through the end of 2011 as shown in Fig. 1. This indicates that there is huge and growing global demand for emissions-free wind power which can be installed quickly and virtually everywhere in the world.

Wind power is usually extracted by wind turbines using either fixed speed or variable speed regimes. The latter is distinctive with getting more energy for a specific wind speed, better aerodynamic efficiency, less mechanical stresses and reduced noise levels. Among of these types, DFIGs represent nowadays the most preferred topology for recent wind farms. The concept of DFIG for variable-speed wind turbine provides the possibility of controlling the active and reactive power, which is significant for grid integration purposes. On the other hand, vector control of the DFIG enables the decoupling between active and reactive power as well as between the torque and the power factor. Hence, unique features of grid supporting are expected. The theoretical background for modeling induction machines is widely developed and is exhaustively dealt with numerous papers and textbooks as seen in the literatures. On the other hand, most of these literatures concerned mainly similar methodologies to the corresponding conventional synchronous ones for acquiring stability studies. However, the dynamic response characteristics of wind turbine generators with either fixed or variable speed operation are different from those conventional synchronous ones. This is mainly due to their operation topologies with their complex power electronics and power exchanges policies. Also, following the requirements for modern grid codes raise extra complexities and various operation modes. These concerns should be considered in order to realize a close behavior of the developed models to the realized ones in the real field. The aim of this chapter is to emphasize the essential requirements for realizing an accurate modeling of grid integrated wind farms for those purposes regarding stability studies. Detailed operation modes of the both SFIG and DFIG generators are explored. Varieties of operation modes of the DFIG one, in particular, are considered including the control of both Rotor Side Converter (RSC) and Grid Side Converter (GSC). Modern rotor protection scenarios are included using either crowbar or chopper circuitries. Moreover, the regulation of recent grid codes are followed. This results in a complete and sophisticated dynamic modeling of wind power plants integrated into grid systems. Different simulation examples are presented using the MATLAB/Simulink as a powerful tool for performing efficient dynamic modeling of complicated power systems. Some simulation examples are presented to corroborate the efficacy of the developed simulation for power system stability studies.

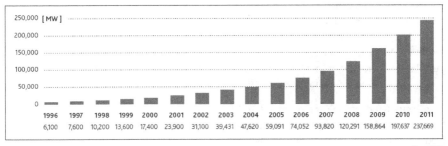

Source: GWEC

Figure 1. Global cumulative installed capacity 1996-2011 (source: GWEC annual report)

2. Wind power plant description

Historically, wind energy had been deployed in small scale and the integration had been carried out at the distribution system level. In this case, the impact of wind turbines generators on power system performance is minimal. In contrast, increasing penetration of wind power plants results in new and different kind of problems than those have been experienced in conventional grids. These problems may be owing to the random nature of the wind and the dynamic characteristics of the wind generators themselves. Fig. 2 shows a schematic diagram of a typical wind farm consisting of (n) units of wind turbines. Nowadays, modern wind farms include 20 to 150 units with typical size from 1.5 MW to 5.0 MW wind turbine generators. Larger sizes up to 7.0 MW are recently available in the market, and they were successfully installed in some European countries. The typical generator's terminal voltage may range from 575 to 690 V with frequency of 50 (or 60) Hz. The generator terminal voltage is stepped up to the Collector Bus system with typical voltage of 22-34.5 kV. The step up transformer is an oil cooled, pad mounted located at the base of the wind turbine unit. Sometimes, the step up transformer is mounted in the turbine nacelle. The typical wind farm collector system consists of a 22-34.5 distribution substation collecting the output of the distributed wind turbine generators through the incoming feeders. Finally, the collected power is transferred to the utility side via an interconnection step up transformer. Usually some reactive power compensation units are provided by a collection of switched capacitors, in addition, certain considerations should be applied for avoiding the harmonic effects [2]. The common type of the wind turbine generators that are commercially available nowadays are induction generator (IG), wound rotor synchronous generator (WRSG), and permanent magnet synchronous generator (PMSG). Due the uncontrollable natural characteristic of wind speed, the induction generators IGs are suitable for driving the wind turbines. The two basic types of wind turbines used nowadays are fixed-speed wind turbine (FSWT) that equipped with squirrel cage SFIGs and variable-speed wind turbines (VSWT) equipped with DFIGs. It is to be noted that, squirrel-cage induction generators work normally within a limited wind speed range, which is one of their main drawbacks in comparison with variable-speed ones. Variable-speed wind turbines are mainly equipped with variable frequency excitation of the rotor circuit, through a partial scale back-to-back converter, whereas the stator windings are connected directly to the AC grid. The main advantage of DFIG wind turbines is their ability to supply power at a constant voltage and frequency while variations of the rotor speed [3]-[6].

Numerous wind turbines are grouped and installed into arrays at one site to build a wind farm of the desired power production capacity. The turbines may be grouped into arrays, feeding power to a utility, with its own transformers, transmission lines and substations. Stand-alone systems catering the needs of smaller communities are also common. A schematic of a typical wind farm consisting of (n) units of wind turbines is demonstrated in Fig. 3. The use of DFIGs in wind farm installations is today a standard practice, due to its suitable characteristics for the wind turbines. Certain considerations should be applied for avoiding the harmonic effects. The typical wind farm collector system consists of a distribution substation collecting the output of the distributed wind turbine generators through the incoming feeders. Usually some reactive power compensation units are provided by a collection of switched capacitors. Finally, the collected power is transferred to the utility side via an interconnection step up transformer. Further details are available in [7].

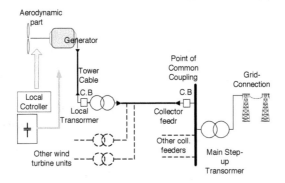

Figure 2. General schematic of a wind farm

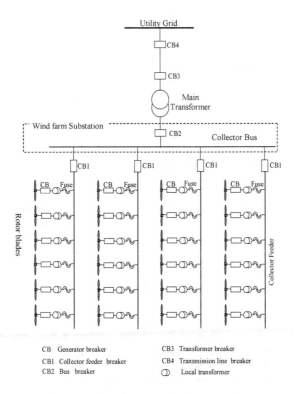

Figure 3. Principle layout of a typical wind farm system.

3. Power system stability overview

"Power system stability" can be generally defined for a power system as, its ability to remain in operation equilibrium under normal operating conditions and to regain an acceptable state of equilibrium after subjecting to a disturbance. Following a large disturbance, if the power system is stable, it will reach a new equilibrium state with practically the entire system intact; the actions of automatic controls and possibly human operators will eventually restore the system to normal state. On the other hand, if the system is unstable, it will result in a run-away or run-down situation, or equivalently a progressive increase in angular separation of generator rotors, or a progressive decrease in bus voltages [8].

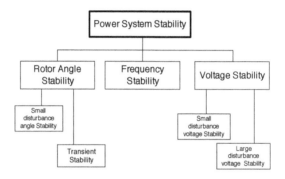

Figure 4. Classification of power system stability

From the point of view of defining and classifying power system stability, there are three important quantities for its operation; angles of nodal voltages, nodal voltage magnitudes, and system frequency. Referring to Fig. 4, the power system stability can be classified into: rotor (or power) angle stability; (ii) frequency stability; and (iii) voltage stability. Hence, different viewpoints are introduced concerning the stability issues raising different types of stability problems including "Rotor angle stability", "Frequency stability" and "Voltage stability". Rotor angle stability is concerned with the system ability to maintain the equilibrium between electromagnetic torque and mechanical torque of each generator in the system. Instability that may result occurs in the form of increasing angular swings of some generators leading to their loss of synchronism with other generators [9]-[12].

Voltage stability is concerned with the ability of a power system to maintain its steady voltage at all buses in the system under normal operating conditions, and after subjecting to a disturbance. Instability that may result occurs in the form of a progressive fall or rise of voltage of some buses. The possible outcome of voltage instability is loss of load in the area where voltages reach unacceptably low values, or a loss of integrity of the power system. The main factor contributing to voltage instability is usually the voltage drop that occurs when active and reactive power flow through inductive reactances associated with the transmission network. This limits the capability of transmission network for power transfer. The power transfer limit is further limited when some of the generators hit their reactive power capability limits. While the most common form of voltage instability is the progressive drop in bus voltages, the possibility of over-voltage instability also may exist and has been experienced at least on one system. It can occur when EHV transmission lines are loaded significantly below surge impedance loading and under-excitation limiters prevent generators and/or synchronous condensers from absorbing the excess reactive power. Under such conditions, transformer tap changers, in their attempt to control load voltage, may cause voltage instability [12]. Frequency stability is concerned with the ability of a power system to maintain steady frequency within a nominal range following a severe system upset, which result in a significant imbalance between generation and load. It depends on the ability to restore balance between system generation and load, with minimum loss of load. It is of importance to note that, severe system upsets generally result in large excursions of frequency, power flows, voltage, and other system variables.

Since the power system is a highly non-linear system, its stability cannot be analyzed as a single problem, and different aspects should be considered to realize reasonable conclusions. On the other hand, the response of the relevant protective elements and their response times can directly affect the stability profile of the system. This complicates the overall stability problem remarkably. Power system transient stability is defined as its ability to maintain its equilibrium state after subjecting to a severe disturbance, such as, short circuit faults.

Due to the non-linearity of power systems, their stability depends on both the initial conditions and the size of a disturbance. Consequently, angle and voltage stability are divided into small- and large-disturbance stability. "Small signal stability" is the ability of the power system to restore its equilibrium state under small disturbances. Such disturbances occur continually due to a loading, or generation, small variations. It is worthy to note that, the classification of power system stability has been based on several considerations to make it convenient for identification of the causes of instability, the application of suitable analysis tools, and the development of corrective measures appropriate for a specific stability problem. Clearly, there is some overlap between the various forms of instability, since as systems fail, more than one form of instability may ultimately emerge. However, a system event should be classified based primarily on the dominant initiating phenomenon, separated into those related primarily with voltage, rotor angle, or frequency. While classification of power system stability is an effective and convenient means to deal with the complexities of the problem, the overall stability of the system should always be kept in mind, and solution of one category should not be at the expense of another.

On the other hand, the wind farm connectivity and the random behavior of wind profile play a major role for characterizing the contribution of wind power plants during faults. Thus the behavior of fault current distribution and consequently the reaction of the protective scheme are obviously affected. moreover, the own dynamic behavior of the induction generator differs remarkably as compared with conventional synchronous ones [13], [14]. Moreover, the continuous wind speed variations and the interaction of the associated power electronics (for DFIG ones) collaborate together for providing the behavior of these machines during fault periods. More sophisticated and well coordinated relaying schemes should be provided to realize the most appropriate protection methodology for wind farm elements [15]-[21]. Insufficient protective elements, non-integrated control scenarios and improper coordination among protective and control strategies may lead to serious problems for large grid-connected wind farms [22]-[27]. As example for these problems is the accident happening in North Germany on November 4, 2006. The UCTA interconnected grid was affected by a serious incident originating from the North German transmission grid that led to power supply disruption for more than 15 million households, splitting a synchronously connected network into three islands (two under-frequency and one over-frequency). After cascading overloads and tripping, two of three large separated systems (Western Island and North-Eastern Island) ended up with a significant amount of wind generation resources. Western Island (under-frequency state): During the incident, about 40% of the wind power units tripped. Moreover, 60% of the wind power units connected to the grid tripped just after the frequency drop (4,142 MW). Wind power units were automatically reconnected to the grid when the conditions of voltage and frequency were in the accepted range. North-Eastern Island (over-frequency state): Significant imbalance in this subsystem caused rapid frequency increase and triggered the necessary primary, standard and emergency control actions of tripping wind generation units sensitive to high frequency values. Tripping these units (estimated value of 6,200 MW) helped to decrease the frequency value during the first few seconds of disturbance [28].

On the other hand, accurate wind resource assessment is essential for realizing a honest profiling of wind energy possibilities aiming to have a successful application of wind resources. As known, an error of 1% in wind speed measurements leads to almost 2% error in energy output [29]-[31]. Thus the importance of such assessment method is essential. Moreover, wind resources are seldom consistent and vary with time of the day, season of the year, height above the ground, type of terrain and from year to year. As such it should be investigated carefully and completely.

4. Wind farm modeling for stability studies

4.1. SFIG wind generator modeling

4.1.1. Induction generator modelling

The squirrel cage induction generator was represented as an asynchronous machine assigned with a negative mechanical torque. Fig. 5 shows the one line diagram of the implemented "d" and "q" equivalent circuits in MATLAB. The electrical part of the machine model was described with a forth-order state space model, whereas its mechanical part was described with a second-order representation. Both representations were mathematically described by the following equations, where the indices (s, r) were assigned for machine stator and rotor respectively. Detailed explanation of this model was clearly outlined in [32]-[35].

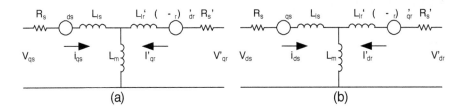

Figure 5. Equivalent circuit of MATLAB-based asynchronous machine a) q-axis equivalent. (b) d-axis equivalent

$$V_{qs} = R_s i_{qs} + \frac{d}{dt}\varphi_{qs} + \omega\varphi_{ds} \tag{1}$$

$$V_{ds} = R_s i_{ds} + \frac{d}{dt}\varphi_{ds} - \omega\varphi_{qs} \tag{2}$$

$$V'_{qr} = R'_r i'_{qr} + \frac{d}{dt}\varphi'_{qr} + (\omega - \omega_r)\varphi'_{ds} \tag{3}$$

$$V'_{dr} = R'_r i'_{dr} + \frac{d}{dt}\varphi'_{dr} - (\omega - \omega_r)\varphi'_{qs} \tag{4}$$

$$T_e = 1.5p(\varphi_{ds}i_{qs} - \varphi_{qs}i_{ds}) \tag{5}$$

where,

$$\varphi_{qs} = L_s i_{qs} + L_m i'_{qr} \tag{6}$$

$$\varphi_{ds} = L_s i_{ds} + L_m i'_{dr} \tag{7}$$

$$\varphi'_{qr} = L'_r i'_{qr} + L_m i'_{qs} \tag{8}$$

$$\varphi'_{dr} = L'_r i'_{dr} + L_m i'_{qs} \tag{9}$$

Also, the mechanical part was described with the following equations.

$$\frac{d}{dt}\omega_m = \frac{1}{2H}(T_e - F\omega_m - T_m) \tag{10}$$

$$\frac{d}{dt}\theta_m = \omega_m \tag{11}$$

4.1.2. Wind turbine modeling

The extracted aerodynamic torque (T_w) is computed as a function of the air density (ϱ), the swept area (A), the wind speed (u) and the power coefficient factor (C_p) as,

$$T_w = \frac{1}{2}\varrho A u^2 \frac{C_p}{\lambda}$$

(12)

Where the tip speed ratio (λ) is expressed as a function of the blade length (R) and the blade angular velocity (ω_b) as,

$$\lambda = \frac{\omega_b R}{u}$$

(13)

The relations among the developed electrical torque (T_g), the mechanical torque (T_m) and the extracted aerodynamic torque (T_w) can be described as functions of the angular velocities of the wind turbine rotor (ω_r) and the generator shaft (ω_g) as,

$$T_w - T_m = J_r \frac{d\omega_r}{dt}$$

(14)

$$T_m = D_{mc}(\omega_r - \omega_g) + K_{mc}\!\int(\omega_r - \omega_g)dt$$

(15)

$$T_m - T_g = J_g \frac{d\omega_g}{dt}$$

(16)

Where the constants J_r, J_g, D_{mc} and K_{mc} are assigned for wind turbine rotor inertia, generator shaft inertia, mechanical coupling damping and mechanical coupling stiffness respectively [36].

4.1.3. Control system modeling

The pitch angle is controlled for limiting the generated power for larger wind speed over the predetermined normal one. The pitch control was accomplished simply with a PI controller, whereas the pitch servo is modeled with a first order delay system with a time constant T_d as shown in Fig. 6.

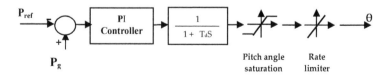

Figure 6. Pitch angle control schematic

4.1.4. Calculation of the FSIG active and reactive powers

Assuming the bus at which the induction generator is connected with the considered power system to be a voltage controlled bus, the load flow computations are carried out to determine the bus injected reactive power Q_{inj} and its voltage angle. Loop current equations are written as

$$V_s \llcorner \delta = -\left(r_s + jX_s + jX_m\right)I_s \llcorner \theta_s + jX_m{}^* I_r \llcorner \theta_r \qquad (17)$$

$$0 = jX_m{}^* I_s \llcorner \theta_s - \left(\frac{r_r}{s} + jX_r + jX_m\right)I_r \llcorner \theta_r \qquad (18)$$

Separating the real and imaginary values in eqns. (17) and (18), it is obtained the following equations,

$$V_s \cos \delta = -I_s\left[r_s \cos \theta_s - \left(X_s + X_m\right)\sin \theta_s\right] - I_r\left[X_m \sin \theta_r\right] \qquad (19)$$

$$V_s \sin \delta = -I_s\left[r_s \sin \theta_s + \left(X_s + X_m\right)\cos \theta_s\right] + I_r\left[X_m \cos \theta_r\right] \qquad (20)$$

$$0 = -I_s\left[X_m \sin \theta_s\right] - I_r\left[\frac{r_r}{s}\cos \theta_r - \left(X_s + X_m\right)\sin \theta_s\right] \qquad (21)$$

$$0 = I_s\left[X_m \cos \theta_s\right] - I_r\left[\frac{r_r}{s}\sin \theta_r + \left(X_s + X_m\right)\cos \theta_s\right] \qquad (22)$$

Then, the generator output active power equation can be written as

$$P_{out} = -\frac{I_r^2}{s}r_r - I_s^2 r_s \qquad (23)$$

Also, the induction generator input reactive power Q_{IG} is computed as

$$Q_{IG} = I_s^2 X_s + I_r^2 X_r + |I_s \llcorner \theta_s - I_r \llcorner \theta_r|^2 X_m \qquad (24)$$

4.2. DFIG wind generator modeling

4.2.1. Mathematical representation of DFIG

The mechanical, electrical power and the slip (s) of the DFIG could be obtained from the equations

$$P_m \approx -P_r \frac{1-s}{s} \approx P_s + P_r, P_r = -sP_s \qquad (25)$$

where:-

$$s = \frac{\omega_s - \omega_r}{\omega_s} \tag{26}$$

Pm, Ps, Pr are the mechanical, stator, and rotor power respectively. ωs, ωr the synchronous and rotor speed respectively.

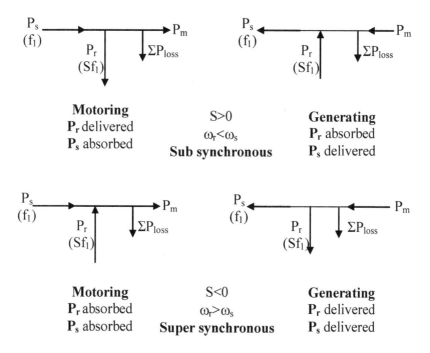

Figure 7. Operation modes of WRIG.

Power exchanging between the DFIG and the grid is complicated depending upon the rotor slip "s" and weather the machine speed is either over or below the synchronous speed. As noticed from Equ.1, the higher the slip, the larger the electric power absorbed (or delivered) through the rotor. DFIG is used in applications with limited speed control range ($|Smax| < 0.2$ to 0.3), so the rating of the rotor-side static converter is around PSN $|Smax|$, where PSN is the rated stator power. Running into the generating (or motoring) mode is determined by the sign of rotor slip "s" and the active power input (or extracted) electrically from the rotor "Pr". As summarized in Fig. 7, the DFIG can operate in the generating or the motoring mode whenever it runs either sub-synchronously ($\omega r < \omega s$) or super- synchronously ($\omega r > \omega s$) [37]. The rotor-side converter operates as a rectifier and stator-side converter operate as an inverter for motoring mode (with sub-synchronous speed) or for generating mode (with super-synchronous speed), whereas the slip power delivered from the rotor. The rotor-side converter operates as an inverter and stator-side converter as a rectifier for generating mode (with sub-synchronous speed) or for the motoring mode (with super-synchronous speed), whereas the slip power is supplied to the rotor. At the synchronous speed, slip power is taken from supply to excite the rotor windings and in this case machine behaves as a synchronous machine [38]. The power flow through the rotor winding depends mainly on the rotor speed. If the rotor runs super-synchronously, both stator and rotor generated powers are fed to the grid, whereas rotor circuitry is fed with active power in the sub-synchronous mode. On the other hand, the reactive power flow can be calculated as,

$$
\begin{aligned}
Q_s + Q_r &= 3 \operatorname{Im} ag(V_s I_s^*) + 3 \operatorname{Im} ag(\frac{V_r I_r^*}{S}) \\
&= 3\omega_s (L_{sl} I_s^2 + L_{rl} I_r^2 + L_m I_m^2)
\end{aligned}
\tag{27}
$$

where

Qs, Qr are stator and rotor reactive power.

Vs, Vr, Is, Ir are the voltage and current of the stator and the rotor respectively.

Lsl, Lrl, Lm are the stator leakage, rotor leakage, and magnetizing inductances respectively.

Hence, the reactive power required to machine excitation may be provided by the rotor, the stator or by both. If the stator operates at unity power factor ($\varphi s = 0$), the rotor-side static power converter has to deliver reactive power extracted either internally (through DC linking capacitor) or absorbed from the grid (over-excited DFIG). For under excitation conditions, on the other hand, unity power factor is assumed in the rotor winding ($\varphi r = 0$) extracting the required excitation from the stator winding.

4.2.2. Principles of vector control of RSC

Vector control aims principally (for induction machines) to perform the control action in order to produce high-dynamic performance similar to those realized with DC machines. To achieve this target, the reference frames may be aligned with the stator flux-linkage space vector, the rotor flux-linkage space vector or the magnetizing space vector. Stator-Flux Oriented vector control approach is deployed for RSC to provide independent control of active and reactive power. In other words, the vector-controlled DFIG enables the decoupling between active and reactive power as well as between the torque and the power factor [38].

The most common approach in dynamic modeling of DFIGs for wind turbines is using a space vector theory based model of a slip-ring induction machines [37].

The space-phasor model of the WRIG, for steady state, in synchronous rotating reference frame, is characterized by DC quantities that make it suitable for control design by transforming into d-q axis using the Park transformation. After transforming from abc to d-q coordinates, the voltage equations of the IM in a general rotating reference frame are calculated from.

$$V_{ds} = \frac{d\Psi_{ds}}{dt} + R_s i_{ds} - \omega_s \Psi_{qs} \quad , \qquad V_{qs} = \frac{d\Psi_{qs}}{dt} + R_s i_{qs} + \omega_s \Psi_{ds} \tag{28}$$

$$V_{dr} = \frac{d\Psi_{dr}}{dt} + R_r i_{dr} - (\omega_s - \omega_r)\Psi_{qr}$$
$$V_{qr} = \frac{d\Psi_{qr}}{dt} + R_r i_{qr} + (\omega_s - \omega_r)\Psi_{dr} . \tag{29}$$

where,

Vds, Vqs, Ids, Iqs are the stator voltages, and currents in d-q reference frame, respectively.

Vdr, Vqr, Idr, Iqr are the rotor voltages, and currents in d-q reference frame, respectively.

ψds, ψqs, ψrs, ψqr are the flux linkages in d-q reference frame for stator and rotor respectively.Rs, Rr are the stator and rotor resistances, respectively.

Flux linkage equations for stator and rotor windings can be calculated as,

$$\Psi_{ds} = L_s I_{ds} + L_m I_{dr} \quad , \qquad \Psi_{qs} = L_s I_{qs} + L_m I_{qr}$$
$$where \qquad L_s = L_{sl} + L_m \tag{30}$$

$$\Psi_{dr} = L_r I_{dr} + L_m I_{ds} \quad , \qquad \Psi_{qr} = L_r I_{qr} + L_m I_{qs}$$
$$where \qquad L_r = L_{rl} + L_m \tag{31}$$

For the rotor-side controller the d-axis of the rotating reference frame used for d-q transformation is aligned with air-gap flux ψ_s. Aligning the system of coordinates to stator flux seems most useful, as, at least for power grid operation, ψ_s is almost constant, because the stator voltages are constant in amplitude, frequency, and phase [37]. Aligning the d-axis of rotating reference frame to the stator flux linkage (stator flux oriented control) will result in,

$$\overline{\Psi}_s = \Psi_s = \Psi_{ds} \quad , \qquad \Psi_{qs} = 0, \qquad \frac{d\Psi_{qs}}{dt} = 0 \tag{32}$$

Since the variation of the stator flux ψds is considered to be very small, $\frac{d\Psi_{ds}}{dt} = 0$.

Then, Eqns. (30) and (31) can be rewritten as,

$$I_{qs} = -\frac{L_m}{L_s} I_{qr} \quad , \quad I_{ds} = \frac{\Psi_{ds} - L_m I_{dr}}{L_s} \tag{33}$$

Then, neglecting the stator resistance (Rs) simplified Eqns. (3) and (4) to be expressed as:,

$$V_{ds} = 0 \quad , \quad V_{qs} = \omega_s \Psi_{ds} \tag{34}$$

The resulting active and reactive power can be calculated as,

$$
\begin{aligned}
P_s &= \frac{3}{2}(V_{ds} I_{ds} + V_{qs} I_{qs}) = \frac{3}{2} V_{qs} I_{qs} = -\frac{3}{2}\frac{L_m}{L_s}\omega_s \Psi_{ds} I_{qr} \\
Q_s &= \frac{3}{2}(V_{qs} I_{ds} - V_{ds} I_{qs}) = \frac{3}{2} V_{qs} I_{ds} = \frac{3}{2L_s}\omega_s \Psi_{ds}(\Psi_{ds} - L_m I_{dr})
\end{aligned}
\tag{35}
$$

Eqn. (35) clearly shows that the active power delivered (or absorbed) under stator flux orientation control by the stator, Ps, may be controlled through the rotor current Iqr, while the reactive power (at least for constant ψs) may be controlled through the rotor current Idr. As pulse-width modification on the machine-side converter is generally performed on rotor voltages, voltage decoupling in the rotor is required. Substituting from Eqn. (33) into Eqn. (31) yields,

$$\Psi_{dr} = L_{sc} I_{dr} + \frac{L_m}{L_s}\Psi_{ds} \quad , \quad \Psi_{qr} = L_{sc} I_{qr}, \quad L_{sc} = L_r - \frac{L_m^2}{L_s} \tag{36}$$

Substituting from Eqn. (11) into Eqn. (4) yields,

$$
\begin{aligned}
V_{dr} &= R_r i_{dr} + L_{sc}\frac{dI_{dr}}{dt} - [s\omega_s L_{sc} I_{qr}] \\
V_{qr} &= R_r i_{qr} + L_{sc}\frac{dI_{qr}}{dt} + \left[s\omega_s\left(\frac{L_m}{L_s}\Psi_{ds} + L_{sc} I_{dr}\right)\right],
\end{aligned}
\tag{37}
$$

$$s\omega_s = \omega_s - \omega_r$$

Eqn. (37) constitutes the rotor voltage decoupling conditions, where the terms in brackets represent the decoupling or compensating terms. Adding these compensating terms to the corresponding uncompensated voltage terms (the outputs of the current controllers which control Isd and Isq components) makes it possible to achieve decoupled performance of the stator flux-oriented control of the rotor-side converter [38], [39]. The overall control scheme, described before, was summarized in Fig. 8.

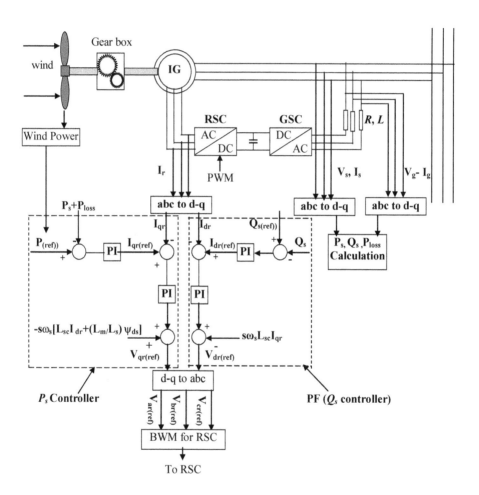

Figure 8. Vector control structural diagram of the RSC for DFIGs.

The reference power, P_s(ref), is calculated from the power speed curve of the wind turbine. The ABCD locus represents the maximum power control action as illustrated in Fig. 9. Each of these curves was structured for a certain wind speed as a function of the air density (ϱ), the turbine blade radius (R), the tip speed ratio (λ), the wind speed (v), the power coefficient (Cp), the pitch angle (β) and the wind turbine speed (ωt) as,

$$P_w = \frac{1}{2}\rho\pi R^2 v^3 C_p(\lambda,\beta). \qquad where \qquad \lambda = \frac{\omega_t R}{v} \qquad (38)$$

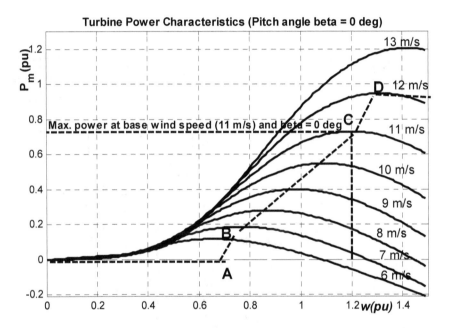

Figure 9. P-ω curve and tracking maximum power curve of the wind turbine.

4.2.3 Principles of vector control of GSC

The main objective of GSC is to maintain the DC-link voltage to be constant regardless of the magnitude and direction of the slip power Pr. As illustrated in Fig. 10, a current-regulated PWM scheme is used, where d and q axes currents are used to regulate DC-link voltage and reactive power. On the other hand, the GSC uses a power filter to reduce current harmonics flow into the power source. The inductance and resistance (R, L) of the input filter are taking into account, so, the voltage equations across the inductor (L, R) can be written as follows:

$$V_g = RI + L\frac{dI}{dt} + V \qquad (39)$$

Where I, V are the current and voltage at the grid side. These equations may be translated into d–q synchronous coordinates that may be aligned with the positive sequence of grid side voltage ($V_{gq} = 0$, $V_{gd} = V_g$):

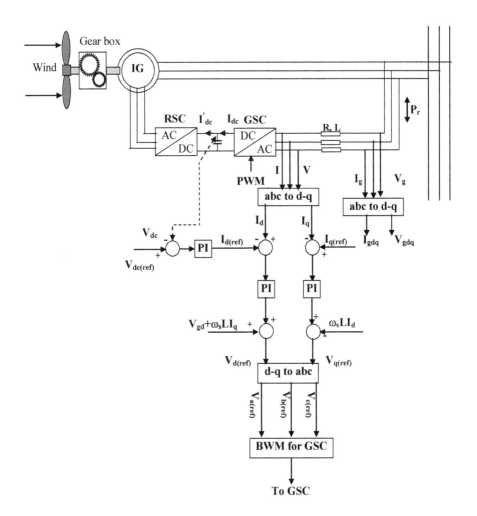

Figure 10. Vector control structural diagram for a GSC of a DFIG.

4.2.4. Grid integration issues of DFIGs

As described in the preceding analysis, the concept of DFIG for variable-speed wind turbine provides the possibility of controlling the active and reactive power, which is significant for grid integration. On the other hand, vector control of the DFIG enables the decoupling between active and reactive power as well as between the torque and the power factor. Hence, unique features of grid supporting are expected [40]-[43]. Owing to the new policies of recent grid codes, wind farms are required to remain grid- connected during grid faults for a certain time so that they can directly contribute with active and reactive power to the grid. This leads to support the overall system stability [44], [45]. On the other hand, wind turbines are separated from the grid following grid faults leading to loss of an undesirable portion of power generation. Hence, most utilities require Fault Ride-Through (FRT) capability for grid-connected wind farms. Theoretically, decoupling between the wind generators and the grid with the utilized back-to-back converters in variable speed machines greatly eliminates impacting the system frequency. However, large penetration of wind energy into the grid leads the spinning reserve to have less time to react to the power imbalance resulting from load variations or wind speed fluctuations. Hence considerable frequency deviations can be expected in the system. Then, grid voltage support by wind generators is essential for eliminating these impacts. Basic control strategies for support the grid was conventionally achieved by controlling the behavior of either the Rotor Side Coveter (RSC) or the Grid Side Converter (GSC) or both of them to maintain the generated active and reactive power of the DFIG during the fault period. Recently, more control loops were proposed utilizing the system frequency response inertial response to do the same action, where the fluctuation in system frequency or apparent system inertia can be utilized to control the profile of the generated power by the wind generators [46]. Also, damping control loops were also proposed to eliminated the expected mechanical oscillations on the DFIG rotor to serve for controlling the generated active and reactive power [41], [47].

Old grid integration scenarios recommend to separate wind generators from the grid when their terminal voltage level decreased below 80% of its nominal range. Since these scenarios are not acceptable with the increasing penetration wind power plants, recent grid codes demanded that grid integrated wind farms must withstand voltage dips to a certain percentage of its nominal voltage (down to 0% in some cases) for a specified duration. Differences among these codes were demonstrated in Fig. 11 [48]-[50]. This is mainly to benefit with its capabilities for supporting the grid with reactive power to avoid the possibilities of voltage instability problems as possible. Accordingly, wind turbines are usually considered as active components facilitating grid support when required.

According to Fig. 12 (German grid code), wind turbines have to be still connected to the grid during the elapsed time by areas I and II even when the voltage at the Point of Common Coupling (PCC) with the grid drops to zero. For severe thermal impacts, on the other hand, wind turbines can be separated from the grid and then resynchronized shortly after a few hundreds of milliseconds [48]. During the fault period, power generation by the variable speed wind generator is reduced by the converter control temporarily. Then, grid supporting is released to increase the generated active power to the grid. Once the fault is cleared, the wind turbine controller should be able to increase the active power fed to the grid after resynchronization process as fast as possible. In area III short disconnection from the grid is allowed if the voltage cannot return to area II. Within the next two seconds, resynchronization is always required. If the voltage remains low longer than 2.683 seconds, tripping of the corresponding wind turbines by system protection is issued [50], [51].

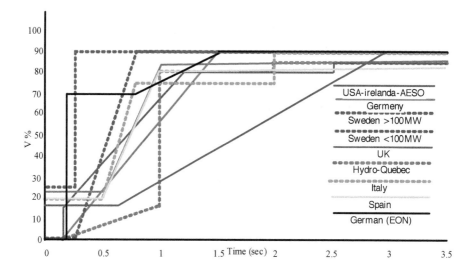

Figure 11. FRT requirements for various grid codes

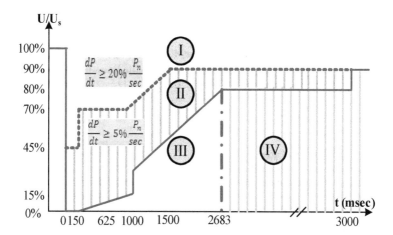

Figure 12. Fault Ride Through capability of wind farm power station

These grid codes assisted different possible voltage control strategies for regulating the terminal voltage DFIG wind turbines during normal/abnormal operation. Principally, DF|IG voltage can be controlled by either the RSC [45] or the GSC [52] or by both of them to support the grid by generating reactive power during network faults and restore the grid voltage as possible. If the PCC voltage is dropped more than 10% of the RMS of the generator terminal voltage, the wind generator must be switched to voltage support mod. As described in Fig. 3, the network voltage support is initiated within 20 ms. by providing reactive power at the generator terminals. A reactive power output of 100% of the rated current must be possible if necessary as well. On the other hand, it is worthy to note that the requirements described in Fig 13 are not applicable when the wind turbine generator output power is less than 5% of its rated power or during very high wind speed conditions [50].

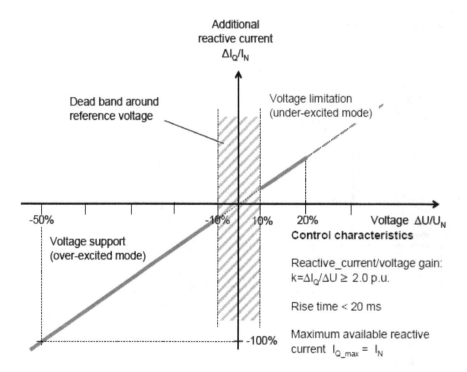

Figure 13. Recommended voltage German support profile

4.2.5. Rotor protection scenarios

As the consequence of a sudden short circuit on the grid, the stator current of the DFIG may contain DC-components which appear in the rotor side as an AC current. Therefore the rotor current can exceed to more than 2-3 times its nominal value [52]-[56]. This obviously is not acceptable, and it can in turn cause large increases in the DC-link voltage or cause damage to the rotor-side converter itself. With the continuous increase in the rotor current or the DC-voltage, the rotor side converter should be disconnected otherwise it will be destructed. So, to satisfy the FRT, keeping the machine connected to the grid without destroying the machine, both of the crowbar and chopper are used in this thesis for the wind farm rotor protection.

The crowbar is a device mounted in parallel with the rotor converter, as sketched in Fig. 14, which comprises a three phase diode bridge that rectifies the rotor currents and a single thyristor in series with a resistor R_{crow}. For the grid short circuit disturbance the crowbar short-circuits the rotor terminal, and hence the RSC is separated from the rotor circuit, and the DFIG goes to operate as a FSIG, in which the DFIG is not controlled by the rotor-side converter. The thyristor of the crowbar circuit is triggered by the DC-voltage, at 1.1 of the nominal DC voltage value, which rises due to rotor current increase. After a short time delay of about 60-120ms during the crowbar is switched off and the DFIG regain its normal operation.

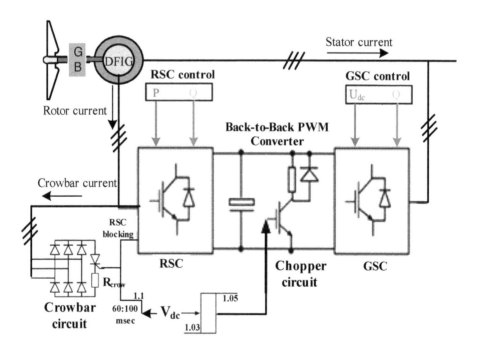

Figure 14. A DFIG equipped with the crowbar and chopper.

According to the situation of operation during the fault, different operation modes occurs affecting the behavior of the DFIG as shown in Fig. 15 as follows [57]-[64]. In *Normal mode* (mode 1), rotor current and rotor voltage are controlled by the IGBT's. In *Crowbar mode* (mode 2), the Rotor side IGBT-converter is switched off, whereas the crowbar is switched on. When the crowbar is on, the rotor side converter controller is stopped and reset in this mode. Then the machine behaves as a SFIG one. In the *No load mode* (mode 3), the rotor side IGBT-converter is switched off and consequently the rotor-current = 0. In *Deactivation mode* (mode 4), the generator rotor windings are fed by anti-parallel diodes of rotor side converter when the IGBT-converter is deactivated. In mode 4, the DFIG can be described by the same equations as used for normal mode. However, the absolute value of rotor voltage is determined by the DC-link voltage only.

All these operation scenarios affect the performance of the DFIG generator during grid faults and consequently influence the stability profile of such generators during stability problems.

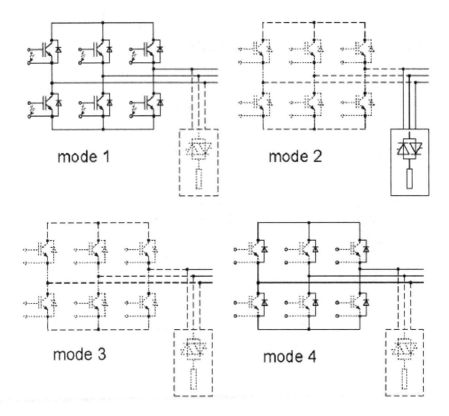

Figure 15. Operation modes with the crowbar operation modes

It is found that, the crowbar ignition converts the DFIG to SFIG by short-circuiting the rotor circuit, and blocking the RSC. During this time the generator acts as a common single fed induction generator and consumes reactive power, which is not desirable. Therefore, utilizing crowbar mechanisms has recently replaced with a DC chopper. The DC chopper is used to limit the DC voltage by short-circuiting the DC circuit through the chopper resistors as demonstrated in Fig. 14.

To keep the DC voltage below the upper threshold, 1.05 of the nominal value, the chopper is switched on by IGBT switches and stay in conduction mode until the DC voltage decreased to the lower threshold value, 1.03 of the nominal value. [65].

When the DC voltage is maintained by the chopper the DFIG can be controlled even if it is operating on a low voltage level. However, in extreme situations the DC voltage may increase further and the crowbar must be used to protect the DFIG from damage.

5. Simulation examples for transient stability studies

5.1. Selected testing network

The well known IEEE standard 3-generator, 9-bus power system, shown in Fig. 16, is considered for preparing the required simulation examples [66]. It is assumed that each of the system generators is equipped with the automatic voltage regulator (AVR) and speed governor as described in Fig. 17 and 18 [67] considering KA=25, TA=0.05, Kg=50 and Tg=5.

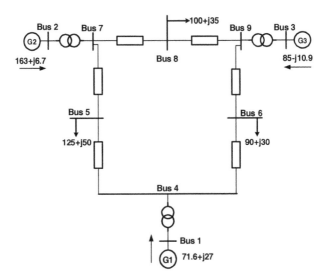

Figure 16. One line diagram of the selected IEEE 9-bus system.

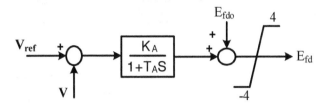

Figure 17. IEEE Type-AC1A Excitation system representation

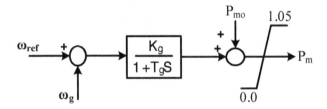

Figure 18. Speed governor system representation

5.2. Simulated examples for SFIGs

In order to emphasize the paper aim, a preliminary stability study was performed with the selected IEEE 9-bus system utilizing the developed simulation platform. Effects of considerable amounts of wind energy conversional systems in conjunction with conventional synchronous generators are highlighted. One of these conventional ones (connected at bus 3) is replaced by a FSIG equivalent generator representing the overall wind farm. Also, the FSIG output active power is assumed to equal 0.85 pu. (the removed generator output power). This power is equal to about 27% of the system total generated power. Referring to Fig. 4 the FSIG loop current equations are solved. It is found that the generator steady state slip So= -0.0181, and its input reactive power equals 0.4512 pu. Hence, after carrying out the system load flow study, it is chosen a capacitor of the reactive power 0.38624 pu. to be connected with the FSIG. Each of the system two synchronous generators was represented by the well known three-axis model, whereas the SFIG was represented using the described equations earlier.

A three phase short circuit fault was initiated at 0.10 sec. at bus 8. It is found that both the system synchronous (angle) and voltage stability are lost, as shown in Figs. 19 and 20. The system can maintain its stability when the fault is removed after elapsing a time interval not larger than 0.225 sec. from the fault instant. As concluded from the results, the system can maintained its both voltage and synchronous stability, if the faulted line is isolated before 0.285 sec. from the fault instant. On the other hand, the original system (equipped with three synchronous generators only) has the critical times for isolating the faults of the first and second locations of 0.25 and 0.3 sec. respectively.

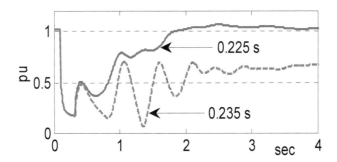

Figure 19. Wind farm terminal voltage for SFIG unit

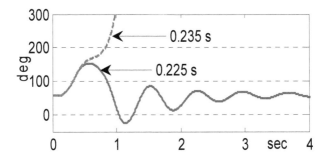

Figure 20. Synchronous machines relative angle δ_{21} for SFIG unit

Another testing example was carried out when the wind speed was increased from 12 to 15 m/sec. whereas a 3-phase short circuit fault was applied at bus 8 after 0.1 sec. Referring to Figs 21, 22 and 23, both the system voltage and frequency stability were maintained when the fault duration time is less than 0.15 sec. These presented simulation examples may raise different conclusions. The critical time value for removing a short circuit at one of the system buses or isolating a faulted line, are about 90% of those for the original power system. This essentially means that replacing a synchronous generator by an FSIG may deteriorate the system stability. For an increase in the wind speed which is followed by a 3-phase short circuit at one of the system buses, the critical time for removing the fault is equal to about 65% of that for the short circuit fault under the normal wind speed. This means that increasing the wind speed deteriorates the system stability.

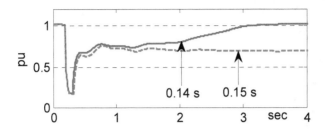

Figure 21. Wind farm terminal voltage for SFIG unit

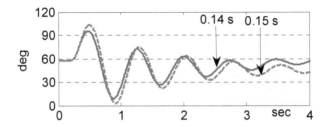

Figure 22. Synchronous machines relative angle δ21 for SFIG unit

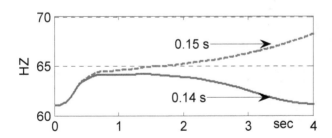

Figure 23. Frequency of the induction generator for SFIG unit

5.3. Simulated examples for DFIGs

It was considered that the SFIG was replaced with a DFIG one. Similar tests were applied as described earlier. Unlike the SFIG one, the FRT as well as the rotor protection schemes affected the performance of the DFIG during faults. The crowbar firing is triggered by the DC-voltage (with a threshold of 110% of its nominal DC voltage) which rises due to the first rotor current peak. The IGBT's are usually stopped by its own protection but the current and thus the energy continues to flow into the DC-link through the freewheeling diodes leading to a very fast voltage increase. Crowbar is typically kept on for 60-120 ms and then it is switched off. Then the converter resynchronizes and the system goes back into controlled DFIG operation. When the crowbar is switched on, the converter is separated from the rotor circuits.

Referring to the developed modeling, Fig. 24 shows the performance of the DFIG during a three phase short circuit while the crowbar circuitry was activated. As a result, rotor, crowbar and stator currents were demonstrated in Fig. 24 (a), (b) and (c) respectively. As soon as the Crowbar DC link voltage exceeded its limit, the crowbar was initiated as shown in Fig. 25(a) and the rotor side converter is blocked. The Crowbar thyrestor firing is switched off after elapsing 100 ms, the DFIG terminal voltage is nearly equal to zero as demonstrated in Fig. 25(b) during the fault period. As noted from Fig. 25(c), there is no output active and reactive power flow during the fault time as well.

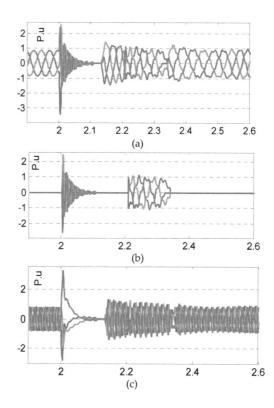

Figure 24. Behavior of the DFIG for a grid fault at bus B6 with crowbar operation (a) Rotor current (b) Crowbar current (c) Stator current

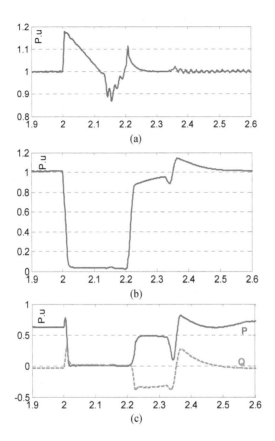

Figure 25. Response of the DFIG for a grid fault at bus B6 with crowbar operation (a) DC link voltage (b) Terminal voltage of DFIG (c) active/reactive power of DFIG

The fault was repeated considering the chopper operation instead of the crowbar. As described in Fig. 26, the chopper circuit was initiated as soon as the DC voltage exceeded 1.05 % of its nominal value and continue until the DC voltage reduced to 1.03. Hence, a discontinuous chopper current was observed as shown in Fig. 26(c). Then, the chopper operation protected the rotor winding while keeping variable speed operation of the DFIG during the fault. Unlike the crowbar operation, this facilitates the grid support capabilities during such faults. As a result, a better terminal voltage of the DFIG was realized as compared with the corresponding crowbar response as described in Fig. 27(d).

As emphasized from the results, the use of the chopper improves the profile of the reactive power injection during and after the fault period. After the fault removing the chopper prevent the operation of the crowbar, as compared with its corresponding operation. Then, the DFIG doesn't absorb a reactive power from the grid. Therefore the chopper protection circuit can support the PCC voltage remarkably. Different tests were applied revealing the perfect capabilities of the DFIGs for supporting the grid during faults as fully described in [62], [65].

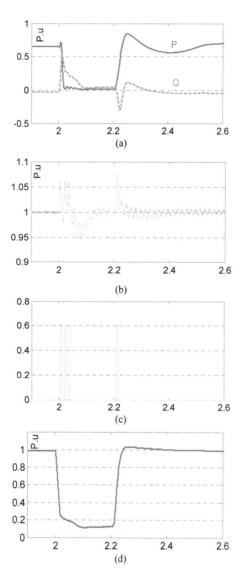

Figure 26. Response of the DFIG for a grid fault at bus B6 with chopper operation (a) Output active and reactive power of DFIG (b) DC link voltage (c) Chopper current (d) Terminal voltage of the DFIG

Author details

Tamer A. Kawady and Ahmed M. Nahhas

Electrical Engineering Department, Umm Al-Qura University, Makkah, Saudi Arabia

References

[1] Global Wind Energy Council, GWEC, "Global Wind Energy: Annual Market Update 2011", GWEC, 2011, www.gwec.org

[2] Tamer Kawady, Naema Mansour and A. I. Taalab, "Wind Farm Protection Systems: State of the Art and Challenges", a book chapter in "Alternative Energy", edited by D. N. Gaonkar, ISBN 978-953-307-046-9, Intechweb, Vienna, Ostria, 2010.

[3] James F. Manwell, Jon G. McGowan, " Wind Energy Explained: Theory, Design and Application",, Wiley; 2 edition (February 9, 2010)

[4] Tony Burton (Author), Nick Jenkins (Author), David Sharpe (Author), Ervin Bossanyi (Author), Wind Energy Handbook, Wiley; 2 edition (June 21, 2011)

[5] Siegfried Heier (Author), Rachel Waddington (Translator), "Grid Integration of Wind Energy Conversion Systems", Wiley; 2 edition (June 5, 2006).

[6] Thomas Ackermann, "Wind Power in Power Systems", Wiley; 2 edition (May 29, 2012).

[7] Mohd. Hasan Ali, "Wind Energy Systems: Solutions for Power Quality and Stabilization ", CRC Press; 1 edition (February 16, 2012).

[8] Jan Machowski, Janusz W. Bialek, James R. Bumby, " Power System Dynamics: Stability and Control", 2008, Second Edition, John Wiely and Sons, LTD.

[9] P.Vijayan and S.Sarkar, and V. Ajjarapu, "A Novel Voltage Stability Assessment Tool to Incorporate Wind Variability", Power & Energy Society General Meeting, 2009. PES '09. IEEE 26-30 July 2009.

[10] Mohsen Rahimi, Mostafa Parniani, "Dynamic behavior and transient stability analysis of fixed speed wind turbines", Renewable Energy 34 (2009) 2613–2624

[11] F. Wua, X.-P. Zhangb, P. Jua, "Small signal stability analysis and control of the wind turbine with the direct-drive permanent magnet generator integrated to the grid", Electric Power Systems Research 79 (2009) 1661–1667.

[12] Libao Shi, Shiqiang. Dai, Yixin Ni, Liangzhong Yao, and Masoud Bazargan, "Transient Stability of Power Systems with High Penetration of DFIG Based Wind Farms", Power & Energy Society General Meeting, 2009. PES '09. IEEE 26-30 July 2009.

[13] Tamer Kawady, Hassan Shaaban and Abdulla El-Said, " Investigation of Grid-Support Capabilities of Doubly Fed Induction Generators During Grid Faults ", accepted for publications in Renewable Power Generation (RPG) - 2011, 6-8 September 2011, Edinburgh, UK.

[14] G. Tsourakisa, B.M. Nomikosb, C.D. Vournasa, "Effect of wind parks with doubly fed asynchronous generators on small-signal stability", Electric Power Systems Research 79 (2009) 190–200.

[15] Durga Gautam, Vijay Vittal, "Impact of DFIG based Wind Turbine Generators on Transient and Small Signal Stability of Power Systems", Power & Energy Society General Meeting, 2009. PES '09. IEEE 26-30 July, 2009.

[16] Eduard Muljadi, C. P. Butterfield, Brian Parsons, and Abraham Ellis, "Effect of Variable Speed Wind Turbine Generator on Stability of a Weak Grid", IEEE Transactions on Energy Conversion, Vol. 22, No. 1, March 2007.

[17] K. A. Folly and S. P. N. Sheetekela, "Impact of Fixed and Variable Speed Wind Generators on the Transient Stability of a Power System Network", Power Systems Conference and Exposition, 2009. PES '09. IEEE/PES 15-18 March 2009.

[18] Durga Gautam, Vijay Vittal, and Terry Harbour, "Impact of Increased Penetration of DFIG-Based Wind Turbine Generators on Transient and Small Signal Stability of Power Systems, IEEE Transactions on Power Systems, Vol. 24, No. 3, August 2009.

[19] Sherif O. Faried, Roy Billinton, "Probabilistic Evaluation of Transient Stability of a Wind Farm", IEEE Transactions on Energy Conversion, Vol. 24, No. 3, September 2009.

[20] F. Wu, X.-P. Zhang, K. Godfrey and P. Ju, "Small signal stability analysis and optimal control of a wind turbine with doubly fed induction generator", IET Gener. Transm. Distrib., 2007, 1, (5), pp. 751–760.

[21] Eknath Vittal, Andrew Keane, and Mark O'Malley, "Varying Penetration Ratios of Wind Turbine Technologies for Voltage and Frequency Stability", Power and Energy Society General Meeting - Conversion and Delivery of Electrical Energy in the 21st Century, 2008,IEEE, 20-24 July 2008.

[22] H.R. Najafi F.V.P. Robinson, F. Dastyar, A.A. Samadi, "Small-Disturbance Voltage Stability of Distribution Systems with Wind Turbine implemented with WRIG", Powereng 2009, Lisbon, Portugal, March 18-20, 2009

[23] E. Muljadi T. B. Nguyen M.A. Pai, "Impact of Wind Power Plants on Voltage and Transient Stability of Power Systems", IEEE Energy 2030 Atlanta, Georgia, USA, 17-18 November 2008.

[24] Ha Thu Le, Surya Santoso, "Analysis of Voltage Stability and Optimal Wind Power Penetration Limits for a Non-radial Network with an Energy Storage System", Power Engineering Society General Meeting, 2007. IEEE 24-28 June 2007.

[25] Nayeem Rahmat Ullah, Torböjrn Thiringer, and Daniel Karlsson, "Voltage and Transient Stability Support by Wind Farms Complying With the E.ON Netz Grid Code", IEEE Transactions on Power Systems, Vol. 22, No. 4, November 2007.

[26] Magni Þ. Pálsson, Trond Toftevaag, Kjetil Uhlen, John Olav Giæver Tande, "Large-scale Wind Power Integration and Voltage Stability Limits in Regional Networks", Power Engineering Society Summer Meeting, 2002 IEEE, Volume 2, 25-25 July 2002 Page(s):762 - 769 vol.2.

[27] E. Vittal, A. Keane, and M. O'Malley," Varying Penetration Ratios of Wind Turbine Technologies for Voltage and Frequency Stability", IEEE-2008.

[28] Union for the Co-ordination of Transmission of Electricity (UCTE), "FINAL Report - System Disturbance on 4 NovembeR 2006", UCTE, Brussels – Belgium, ww.ucte.org.

[29] Tennis M.W., Clemmer S., and Howland J., Assessing Wind Resources: A Guide for Landowners, Project Developers, and Power Suppliers, Union of Concerned Scientist.

[30] Michael C.B., Patrick H., and Rich S., A GIS Assisted Approach to Wide Area Wind Resource Assessment and Site Selection for the State of Colorado, Presented at Windpower 1996, The Annual Conference and Exhibition of the American wind Energy Association, Denever, Colorado, June 23-27, (1996).

[31] Potts J.R., Pierson S.W., Mathisen P.P., Hamel J.R., and Babau V.C., Wind Energy Assessment of Western and Central Massachusetts, AIAA-2001-0060, 2001.

[32] Srinivas R. Chellapilla and B. H, Chowdhury, "A Dynamic model of Induction Generators for Wind Power Studies", Power Engineering Society General Meeting, 2003, IEEE, Volume 4, 13-17 July 2003, pp. 2340-2344

[33] T. Petru and T. Thiringer, "Modeling of wind turbines for power system studies", Power Systems, IEEE Transactions on

[34] Volume 17, Issue 4, Nov. 2002 Page(s):1132 – 1139.

[35] Krause, P.C., O. Wasynczuk, and S.D. Sudhoff, "Analysis of Electric Machinery", IEEE Press, 1995.

[36] Luis M. Fernandez, Jose Ramo Saenz, Francisco Jurado "Dynamic models of Wind farms with fixed speed wind turbines", Renewable energy 31 (2006), Elsevier, pp. 1203-1230.

[37] ION BOLDEA, Polytechnical Institute Timisoara, R man,"Variable Speed Generators". Published in 2006 by CRC Press Taylor & Francis Group.

[38] S. K Salman and Babak Badrzadeh School of Engineering, The Robert Gordon University, Schoolhill, Aberdeen, AB10 1FR, Scotland, U.K. "New Approach for modeling Doubly-Fed Induction Generator (DFIG) for grid-connection studies"

[39] Jaroslav Lepka, Petr Stek, " 3-Phase AC Induction Motor Vector Control Using a 56F80x, 56F8100 or 56F8300 Device Design of Motor Control Application". Freescale Semiconductor Application Note, 2004.

[40] Ekanayake, J.B., Holdsworth, L., Wu, X.G., Jenkins, N. 2003b. "Dynamic Modeling of Doubly Fed Induction Generator Wind Turbines", IEEE Transaction on Power Systems, Vol. 18, Issue 2, May 2003, pp. 803-809.

[41] Anca D. Hansen and Gabriele Michalke, "Voltage grid support of DFIG wind turbines during grid faults", EWEC2007 International Conference.

[42] R. Bena, J. Clare and G. Asher, "Doubly fed induction generator using back-to-back PWM converters and its application to variable speed wind-energy generation", Generation, Transmission and Distribution, IEE Proceedings, Volume 143, Issue 3, May. 1996.

[43] P. Cartwright, L. Holdsworth, J.B. Ekanayake and N. Jenkins, "Coordinated voltage control strategy for a doubly fed induction generator (DFIG)-based wind farm", Generation, Transmission and Distribution, IEE Proceedings, Volume 151, Issue 4, July 1996.

[44] Hansen A.D., Michalke G., Sørensen P., Lund T., Iov F. Co-ordinated voltage control of DFIG wind turbines in uninterrupted operation during grid faults, Wind Energy, No. 10, 2007, pp. 51-68.

[45] Akhmatov V., Variable-speed wind turbines with doubly fed induction generators. Part II: Power System Stability. Wind Engineering, Vol. 26, No. 3, 2002, pp 171-188.

[46] Johan Morren, Jan Pierik and Sjoerd W.H. Haan, "Inertial response of variable speed wind turbines", Electric Power Systems Research, 76 (2006), pp. 980–987.

[47] A. Tapia, G. Tapia, J.X. Ostolaza, "Reactive power control of windfarms for voltage control applications", Renewable Energy, 29 (2004), pp. 377–392.

[48] Zhang Yong, Duan Zhengang, Liu Xuelian, "Comparison of Grid Code Requirements with Wind Turbine in China and Europe", Power and Energy Engineering Conference (APPEEC), 2010 Asia-Pacific, 28-31 March 2010.

[49] I. Erlich, M. Wilch and C. Feltes, "Reactive Power Generation by DFIG Based Wind Farms with AC Grid Connection", Power Tech 2007, IEEE Lausanne, 1-5 July 2007.

[50] I. Erlich, U. Bachmann, "Grid Code Requirements Concerning Connection and Operation of Wind Turbines in Germany", IEEE Power Engineering Society General Meeting, 2005.

[51] Choudhury S., Mohanty K.B., Debta B.K., "Investigation on performance of Doubly-fed induction generator driven by wind turbine under grid voltage fluctuation", Environment and Electrical Engineering (EEEIC), 2011 10th International Conference.

[52] I. Erlich, H. Wrede, and C. Feltes,"Dynamic Behavior of DFIG-Based Wind Turbines during Grid Faults", Power Conversion Conference - Nagoya, 2007. PCC '07.

[53] Tamer Kawady, N. Mansour, A. Osheiba, A. E. Taalab and R. Ramakumar, "Modeling and Simulation aspects of wind farms for protection applications," Proceedings of 40th Annual Frontiers of Power Conference, Oklahoma State University, Stillwater, October 29-30, 2007, pp X-1 to X-7.

[54] Tamer Kawady, "An Interactive Simulation of Grid-Connected DFIG Units for Protective Relaying Studies", IEEE PES/IAS Sustainable Alternative Energy Conference-2009, Valencia, Spain, 28-30 Sept., 2009.

[55] Ekanayake, J.B., Holdsworth, L., Wu, X.G., Jenkins, N. 2003b. "Dynamic Modeling of Doubly Fed Induction Generator Wind Turbines", IEEE Transaction on Power Systems, Vol. 18, Issue 2, May 2003, pp. 803-809.

[56] Anca D. Hansen and Gabriele Michalke, "Voltage grid support of DFIG wind turbines during grid faults", EWEC2007 International Conference.

[57] I. Erlich, M. Wilch and C. Feltes, "Reactive Power Generation by DFIG Based Wind Farms with AC Grid Connection", Power Tech 2007, IEEE Lausanne, 1-5 July 2007.

[58] I. Erlich, U. Bachmann, "Grid Code Requirements Concerning Connection and Operation of Wind Turbines in Germany", IEEE Power Engineering Society General Meeting, 2005.

[59] I. Erlich, W. Winter, A. Dittric, "Advanced Grid Requirements for the Integration of Wind Turbines into the German Transmission System", Power Engineering Society General Meeting, 18-22 June 2006, Montreal, Canada.

[60] Tamer Kawady, N. Mansour, A. Osheiba, and A. I. Taalab, " Detailed Modeling and Control of DFIG Units for Large Wind Farms Using MATLAB ", Middle East Power System Conf. (MEPCON 2009), 20-23 Dec., 2009, Assuit, Egypt.

[61] Wilch, M.; Pappala, V.S.; Singh, S.N. & Erlich, I., " Reactive Power Generation by DFIG Based Wind Farms with AC Grid Connection", IEEE Powertech 2007, 1-5 July 2007, Lausanne, Switzerland, pp. 626-632.

[62] Hassan Shaaban, Tamer Kawady and Abdulla Elsherif, "Transient Stability studies of Power Systems with High Penetration of FSIG-Based Wind farms", Engineering Journal of Minoufiya University, Shebin El-Kom, Egypt.

[63] Istvan Erlich, Jörg Kretschmann, Jens Fortmann, Stephan Mueller-Engelhardt, and Holger Wrede, " Modeling of Wind Turbines Based on Doubly-Fed Induction Generators for Power System Stability Studies", IEEE Transactions on Power Systems, Vol. 22, No. 3, August 2007, pp. 909-919.

[64] K.E. Okedu, S. M. Muyeen, R. Takahashi and J. Tamura, "Wind Farms Fault Ride Through using DFIG with New Protection Scheme", IEEE Trans. on Sustainable Energy, Vol. 3, No. 2, pp. 242 – 254, 2012.

[65] Tamer Kawady, Hassan Shaaban and Abdulla El-Said, " Investigation of Grid-Support Capabilities of Doubly Fed Induction Generators During Grid Faults ", accepted for publications in Renewable Power Generation (RPG) - 2011, 6-8 September 2011, Edinburgh, UK.

[66] P. M. Anderson and A. A. Fouad, "Power System Control and Stability", Text Book, John Wiley and Sons, New York, 2003.

[67] S.M. Muyeen, Junji Tamura and Toshiaki Murata, " stability augmentation of a grid – connected wind farm", Text Book, Springer-Verlag London Limited, 2009.

Chapter 8

Impacts of Wind Farms on Power System Stability

Ahmed G. Abo-Khalil

Additional information is available at the end of the chapter

1. Introduction

Power systems are complex systems that evolve over years in response to economic growth and continuously increasing power demand. With growing populations and the industrialization of the developing world, more energy is required to satisfy the basic needs and to attain improved standards of human welfare. In order to make energy economically available with reduced carbon emission using renewable energy sources, the structure of the modern power system has become highly complex [1].

Nowadays, there are many thousands of wind turbines operating with a total nameplate capacity of 238,351 MW. Between 2000 and 2006, world wind generation capacity quadrupled. The United States pioneered wind farms and led the world in installed capacity in the 1980s and 1990s. In 1997 Germany, as for installed capacity, surpassed the U.S. until once again overtaken by the U.S. in 2008. China has been rapidly expanding its wind installations since the late 2000s and passed the U.S. in 2010 to become the world leader [2].

At the end of 2011, worldwide nameplate capacity of wind-powered generators was 238 gigawatts (GW), growing by 41 GW over the preceding year. Data from the World Wind Energy Association, an industry organization, states that wind power now has the capacity to generate 430 TWh annually, which is about 2.5% of worldwide electricity usage. Between 2005 and 2010 the average annual growth in new installations was 27.6 percent. Wind power market penetration is expected to reach 3.35 percent by 2013 and 8 percent by 2018. Several countries have already achieved relatively high levels of penetration, such as 28% of stationary (grid) electricity production in Denmark (2011), 19% in Portugal (2011), 16% in Spain (2011), 14% in Ireland (2010) and 8% in Germany (2011). At the end of 2011, 83 countries around the world were using wind power on a commercial basis [3].

Europe accounted for 48% of the world total wind power generation capacity in 2009. In 2010, Spain became Europe's leading producer of wind energy, achieving 42,976 GWh. Germany held the top spot in Europe in terms of installed capacity, with a total of 27,215 MW on 31 December 2010 [4].

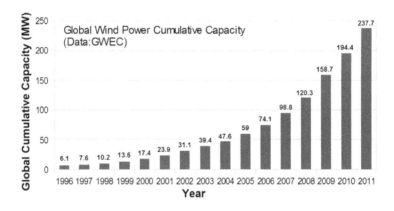

Figure 1. Global Wind Power Cumulative Installed Capacity [2]

The annual energy production of a wind farm is not equal to the sum of the generator nameplate ratings multiplied by the total hours in a year since the wind speed is variable. The capacity factor of a wind farm is the ratio of actual productivity in a year to the theoretical maximum. The range of the capacity factor is between 20 and 40%, with values at the upper end of the range in particularly favorable sites.

The capacity factor is affected by several parameters such as the variability of the wind at the site and the generator size. A small generator would be cheaper and achieve a higher capacity factor but would produce less electricity (and thus less profit) in high winds. On the other hand, large generators would cost more but generate little extra power and may stall out at low wind speed. Thus, the wind farm's optimum capacity factor aimed for would be around 20–35% [6].

Electricity generated from wind power can be highly variable at several different timescales: hourly, daily, or seasonally. However, wind is always in constant supply somewhere, making it a dependable source of energy because it will never expire or become extinct. Annual variation also exists, but is not so significant. Like other electricity sources, wind energy must be scheduled. Wind power forecasting methods are used, but predictability of wind plant output remains low for short-term operation. Because instantaneous electrical generation and consumption must remain in balance to maintain grid stability, this variability can present substantial challenges to incorporating large amounts of wind power into a grid system. Intermittency and the non-dispatchable nature of wind energy production can raise costs for regulation, incremental operating reserve, and (at high penetration levels) could require an increase in the already existing energy demand management, load shedding, storage solutions or system interconnection with HVDC cables. At low levels of wind penetration, fluctuations in load and allowance for failure of large generating units require reserve capacity that can also compensate for variability of wind generation. Thus, integrating significant amounts of wind generation presents a unique challenge to the power system, requiring additional flexibility while simultaneously imposing a decreased capacity factor on conventional generating units [6].

This work investigates the possible impacts of wind power variability, wind farm control strategy, wind energy penetration level, wind farm location, wind intermittent and variability, and wind power prediction accuracy on the power system stability, reliability and efficiency.

2. Power system connection issues of wind farms

Unlike classical sources of energy, wind farms supply real power variations into the upstream grid, and at the same time, in some types of wind generation systems, the reactive power consumption is related to the real power production. These power variations cause voltage variations with consequences for the electrical power system and the customers. On the other hand, the increasing use of power electronics in wind generation systems introduces voltages and current harmonics into the power system. As wind energy is a non-controllable energy source, it can cause problems with voltage stability and transient stability. Due to the rapid increase in the number of wind farms connected to the grid, the increasing rate of power of single wind farm and the weakness of the upstream power grid, where the wind farm connects, the importance and necessity of the study of wind farms connected to power systems is clear.

The connection of wind farm to electrical power systems influences the system operation point, the load flow of real and reactive power, nodal voltages and power losses. At the same time, wind power generation has characteristics with a wide spectrum of influence [4]:

2.1. Location of the wind farm in the power system

The impact of wind farm on the power system depends on the location of wind power plants relative to the load, and the correlation between wind power production and load consumption. Wind power, like any load or generation, affects the power flow in the network and may even change the power flow direction in parts of the network. The changes in the use of the power lines can bring about power losses or benefits. Increasing wind power production can affect bottleneck situations. Depending on its location, wind power may, at its best, reduce bottlenecks, but at another location result in more frequent bottlenecks. There are a variety of means to maximize the use of existing transmission lines such as the use of online information, FACTS, and wind power-plant output control. However, grid reinforcement may be necessary to maintain transmission adequacy and security. Grid extensions are commonly needed if new generation is installed in weak grids far from load centers to make full use of the wind power. The issue is generally the same for modern wind power plants or any other power plants. The cost of grid reinforcements, due to wind power, is therefore very dependent on where the wind power plants are located relative to the load and grid infrastructure, and one must expect numbers to vary from country to country. With current technology, wind power plants can be designed to meet industry expectations such as riding through voltage dips, supplying reactive power to the system, controlling terminal voltage, and participating in SCADA (supervision control and data acquisition) system operation with output and ramp rate control [7, 8, 9].

2.2. Impact of different technologies of wind turbine generator systems

There are many different generator types for wind power applications in use today. The main distinction can be made between fixed speed and variable speed wind generator types.

2.2.1. Fixed speed wind turbine generator

In the early stage of wind power development, most wind farms were equipped with fixed speed wind turbines and induction generators. A fixed speed wind generator is usually equipped with a squirrel cage induction generator whose speed variations are limited. Power can only be controlled through pitch angle variations. Because the efficiency of wind turbines depends on the tip-speed ratio, the power of a fixed speed wind generator varies directly with the wind speed. Since induction machines have no reactive power control capabilities, fixed or variable power factor correction systems are usually required for compensating the reactive power demand of the generator. Fig. 2 shows the schematic diagram of the fixed speed induction machine.

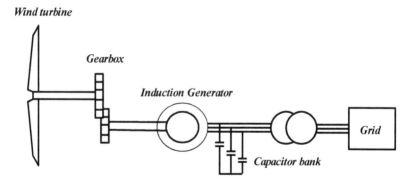

Figure 2. Fixed speed induction generator

2.2.2. Variable speed wind turbine generator

Variable speed concepts allow operating the wind turbine at the optimum tip-speed ratio and hence at the optimum power coefficient for a wide wind speed range. The two most widely used variable speed wind generator concepts are the DFIG and the converter driven synchronous generator.

2.2.2.1. Doubly fed induction generator wind turbine

Due to advantages such as high energy efficiency and controllability, the variable speed wind turbine using DFIG is getting more attention. DFIG is basically a standard, wound rotor induction generator with a voltage source converter connected to the slip-rings of the rotor. The stator winding is coupled directly to the grid and the rotor winding is connected to power converter as shown in Fig. 3.

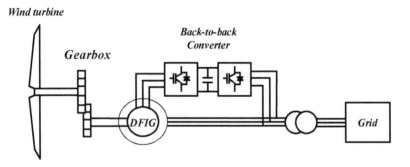

Figure 3. Doubly fed induction generator

The converter system enables two way transfer of power. The grid side converter provides a dc supply to the rotor side converter that produces a variable frequency three phase supply to generator rotor via slip rings. The variable voltage into the rotor at slip frequency enables variable speed operation. Manipulation of the rotor voltage permits the control of the generator operating conditions. In case of low wind speeds, the drop in rotor speed may lead the generator into a sub synchronous operating mode. During this mode, DFIG rotor absorbs power from the grid.

On the other hand, during high wind speed, the DFIG wind turbine running at super synchronous speed will deliver power from the rotor through the converters to the network. Hence, the rotational speed of the DFIG determines whether the power is delivered to the grid through the stator only or through the stator and rotor. Power delivered by the rotor and stator is given by [my papers]:

$$P_R = sP_S \qquad (1)$$

$$P_G = (1 \pm s)P_S \qquad (2)$$

Where, P_G is the mechanical power delivered by the generator, P_S is the power delivered by the stator, and P_R is the power delivered to the rotor.

However, under all operating situations, the frequency of rotor supply is controlled so that, under steady conditions, the combined speed of the rotor plus the rotational speed of the rotor flux vector matches that of the synchronously rotating stator flux vector fixed by the network frequency. Hence, the power could be supplied to the grid through the stator in all the three modes of operation, namely, sub synchronous, synchronous and super- synchronous modes. This provides DFIG a unique feature beyond the conventional induction generator as the latter can deliver power to the grid during super synchronous speed only.

2.2.2.2. Converter driven synchronous generator

This category of wind turbines uses a synchronous generator that can either be an electrically excited synchronous generator or a permanent magnet machine. To enable variable-speed operation, the synchronous generator is connected to the network through a variable frequency converter, which completely decouples the generator from the network. The electrical frequency of the generator may vary as the wind speed changes, while the network frequency remains unchanged. The rating of the power converter in this wind turbine corresponds to the rated power of the generator plus losses. The schematic diagram of the converter driven synchronous generator is as shown in Fig. 4.

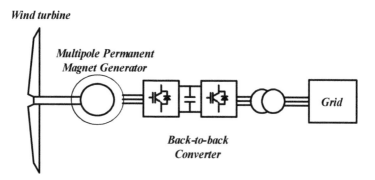

Figure 4. Converter-driven generator

The comparison between the fixed speed and variable speed wind turbines shows that variable speed operation of wind turbines presents certain advantages over constant speed operation. Variable speed wind turbines feature higher energy yields and lower power fluctuations than fixed speed wind turbines. The last feature is particularly important as flicker may become a limitation to wind generation on power systems. Also, variable speed wind turbines produce more reduced loads in the mechanical parts than fixed speed wind turbines. When comparing torque mode control and speed mode control strategies, literature review shows that speed mode control strategy follows wind speed, in order to achieve maximum power coefficient, more accurately, and the higher the speed control loop bandwidth is, the better the tracking is. Nevertheless, as a consequence, it produces more power fluctuations, since speed is rigidly imposed to the turbine. So, from power quality point of view, torque mode control strategy presents better behavior because speed is not directly imposed to the turbine and this control strategy lets the wind turbine to freely change rotational speed during the transient.

2.3. Impacts of wind intermittent and variability

Uncertainty and variability are characteristics that exist in wind power, aggregate electric demand and supply resources and have always posed challenges for power system operators. Future expansion of the loads cannot be predicted accurately, generator outputs and loads fluctuate strongly in different time frames, and it can also lose energy system equipment at any time and without prior warning. Different amounts and types of operating reserves are secured by power system operators to compensate for uncertainty and variability for load reliable service and to keep the system frequency stable. There are many different terms, definitions, and rules concerning what operating reserves entail. The real power capability that can be given or taken in the operating timeframe to assist in generation and load balance and frequency control is defined as the operating reserves. To provide voltage support systems also require reactive power reserve as well, and require certain targets for installed capacity that is often referred to as planning reserve.

The type of event the operating reserves respond to, the timescale of the response and the direction (upward or downward) of the response can differentiate the types of operating reserves. Unpredictable imbalances between load and generation caused by sudden outages of generating units, errors in load forecasting or unexpected deviations by generating units from their production schedules can be compensated by spinning reserve (SR). It becomes more difficult to predict accurately the total amount of power injected by all generators into the power system, as the proportion of power produced by wind farms increases. This added uncertainty must be taken into account when setting the requirement for SR. The uncertainty on the wind power generation increases the uncertainty on the net demand that must be met by traditional forms of generation if wind power generation is considered as a negative load. Spinning reserve is intended to protect the system against unforeseen events such as genera-tion outages, sudden load changes or a combination of both by taking the increased uncertainty into account when determining the requirements for SR.

It is therefore expected that a large penetration of wind generation might require a significant increase in the requirement for SR. However this is not always the case. The cost of SR is indeed far from being negligible. A large number of conventional generating units will need to be synchronized when large amounts of SR must be scheduled for a higher wind-power pene-tration. Therefore, the system operating cost would increase to such an extent that it might be economically desirable to curb this increase in the SR requirement. Determining the optimal amount of SR that must be provided as a function of the system conditions is thus an important and timely issue. The optimal amount of SR is defined as the equality of the cost of generating extra MW of reserve to the benefit that this MW provides, where this benefit is determined as a function of the reduction in the expected cost of interruptions. The ideal case is that the energy and SR amounts and repartitions should be optimized simultaneously. The main difficulties in solving such a problem are: the stochastic nature of the net demand due to the demand and wind forecast errors, and the fact that there are no direct means of incorporating the discrete capacity outage probability distribution in the optimization procedure. The stochastic and highly combinatorial nature of the problem led some researchers to find alternative solutions to the problem [10].

The SR procurement depends on the system as mentioned in [10]. The power system operating cost can increase with the SR provision even though the wind generation reduces the overall net demand. It is also suggested that the extra amounts of MW for reserve can be determined using probabilistic methods combining the uncertain load and wind fluctuations and even including the contingency SR requirements.

2.4. Voltage stability

Power system transient stability is related to the ability to maintain synchronism when subjected to a severe disturbance, such as a short circuit on the bus. System stability is largely associated with power system faults in a network such as tripping of transmission lines, loss of production capacity (generator unit failure) and short circuits. These failures disrupt the balance of power (active and reactive) and change the power flow. Though the capacity of the operating generators may be adequate, large voltage drops may occur suddenly. The unbalance and re-distribution of real and reactive power in the network may force the voltage to vary beyond the boundary of stability. A period of low voltage (brownout) may occur and possibly be followed by a complete loss of power (blackout).

Many of power system faults are cleared by the relay protection of the transmission system either by disconnection or by disconnection and fast reclosure. In all the situations the result is a short period with low or no voltage followed by a period when the voltage returns. A wind farm nearby will see this event. In early days of the development of wind energy, only a few wind turbines were connected to the grid. In this situation, when a fault somewhere in the lines caused the voltage at the wind turbine to drop, the wind turbine was simply disconnected from the grid and was reconnected when the fault was cleared and the voltage returned to normal.

Because the penetration of wind power in the early days was low, the sudden disconnection of a wind turbine or even a wind farm from the grid did not cause a significant impact on the stability of the power system. With the increasing penetration of wind energy, the contribution of power generated by a wind farm can be significant. If the entire wind farm is suddenly disconnected at full generation, the system will loss further production capability. Unless the remaining operating power plants have enough "spinning reserve", to replace the loss within very short time, a large frequency and voltage drop will occur and possibly followed by complete loss of power. Therefore, the new generation of wind turbines is required to be able to "ride through" during disturbances and faults to avoid total disconnection from the grid. In order to keep system stability, it is necessary to ensure that the wind turbine restores normal operation in an appropriate way and within appropriate time. This could have different focuses in different types of wind turbine technologies, and may include supporting the system voltage with reactive power compensation devices, such as interface power electronics, SVC, STATCOM and keeping the generator at appropriate speed by regulating the power etc. [12], [13].

2.5. Impacts of wind farms on power quality

The Danish Wind Industry Association defines power quality as voltage and frequency stability, together with absence of various forms of electrical noise, such as flicker or harmonic distortion. In a power supply system, voltage and frequency must be maintained near nominal values since electrical appliances are manufactured to work under the given alternate current (AC) specification. Conventional power plants fulfill two main tasks in large-scale electrical power systems: power generation and voltage control. In other words, as well as generating power for electricity consumption, they maintain power quality.

For example, at a diesel plant a voltage drop can be countered by simultaneously raising inductance and the steam input to the synchronous generator. The resulting surge in reactive power restores voltage to the desired level. In fact all voltage control devices are effectively controllable reactive power sources. The flexible operation levels of thermal plants allows for a continuous control of reactive power. Feeding intermittent power into electricity grids can affect power quality. The impact depends primarily on the degree to which the intermittent source contributes to instantaneous load (i.e. on power penetration). At low penetrations, wind farms can be connected to the grid as active power generators, with control tasks concentrated at conventional plants. Many studies agree that penetrations of up to 10–20% can be absorbed in electricity networks without adversely affecting power quality and needing extra reserve capacity. Key problems identified at higher penetrations are: At wind speeds below cut-in or above furl-out, wind turbines are disconnected from the grid and left idle. When the wind speed returns to operating range, the turbines are reconnected. The sudden connection of a large turbine can result in brownout (voltage drop incurred when instantaneous load exceeds generated power) due to the current required to magnetize the generator, often followed by a power peak when active power from the generator is fed to the network. There may be times when wind power output exceeds consumer load, making voltage raise above the grid threshold. Cutting off turbines to avoid the excess is not ideal in view of the reconnection problems, but also because ultimately it implies unnecessary shedding of wind energy.

The short-term wind power variations cause voltage fluctuations in the grid, known as flicker because of their effect on light bulbs. Rapid voltage fluctuations can damage sensitive electrical equipment. In a very weak grid, even a single turbine may produce flicker.

Harmonics produced by consumers' electronic equipment can be magnified by wind turbine operation. And more generally, the response of wind farms to an electrical fault may cause transient instabilities which cannot be countered by the control units in the grid. These problems have been reported mainly with reference to small-scale autonomous systems when significant wind power (>100 kW) is connected to a low voltage grid. Stronger grids, with a larger cross-section, have low impedance and so power variations result in smaller voltage variations. However, a sufficiently large wind farm is likely to disrupt power quality even if connected to a high-voltage transmission line [14].

2.6. Reactive power control and voltage control

It is required as a minimum that the reactive power from a large wind farm can be controlled to a specific interval, which is close to unity power factor. However, most wind turbines are also able to provide more advanced reactive power control, which can be useful as grid support. Depending on the technology and the electrical design, such wind turbines will normally have some additional capacity for reactive power, although the available reactive power normally depends on the active power as it does for any other generating units in the power system. This dependency is expressed in the PQ diagrams. The TSO should have access to the reactive power, and the PQ diagram of the wind farm should be delivered by the owner. The additional reactive power capacity can either be used to control constant reactive power or constant power factor, or it can be used in automatic voltage control. In the latter case, it is essential, that it is the voltage in the wind farm point of common coupling (PCC) which is controlled, and that this is done on the wind farm controller level. If the wind turbines are individually attempting to control the voltage in the individual connection points, there is a risk of instability and/or unnecessarily high flow of reactive power between the wind turbines. The possible voltage control in the PCC is of course limited by the limited reactive power available in the wind turbines or from other compensation equipment in the wind farm [15].

2.7. Impacts of wind farm control strategies

One of the challenges which has gained significant importance within the field of electrical power systems over the last years is reactive power control and voltage support from wind farms. Previously the voltage control in the transmission systems was mainly carried out by adjusting the reactive power production or absorption of central power plants, but as the amount of wind power is growing, the requirements for system services including voltage control delivered by wind turbines, and large wind farms in particular, are rising.

So far reactive power control by wind farms has mainly been carried out by utilizing the reactive power capabilities of the wind turbines, but this strategy may not be the most feasible solution when taking into account the new grid code requirements. The optimal reactive power control strategy is influenced by factors like the reactive power capabilities of the wind farm, the on load tap changers of transformers and possible implementation of compensating devices. From a wind farm operator point of view, reactive power control strategies should be based on economic optimization and hence there is a need for investigation of the implications of the new grid code requirements on the reactive power control strategy [16], [17].

2.8. Inertial response with wind

The inertia of traditional synchronous generators plays a significant role in maintaining the stability of the power system during a transient scenario. The inertia dictates how large the frequency deviations would be due to a sudden change in the generation and load power balance, and influences the eigenvalues and vectors that determine the stability and mode shape of transient response. The larger the inertia, the smaller will be the rate of change in rotor speed of the generator during an imbalance in power. This type of response of the traditional synchronous generators is called inertial response. This is a synchronous machine's "reaction," inherently dictated by rotational Newton's law, to sudden changes in the balance between applied mechanical shaft power and electrical power extracted at the generator terminals.

The rate of change of frequency depends on the shortfall or the surplus of generation and the power system inertia. For a given generation shortfall, the higher the system inertia, the lower the rate of change of frequency. Consequently, this inertial response is a critical factor that allows enough time for governor primary control to supply sufficient energy to stabilize system frequency. It is relevant to note that interruptible loads are used to arrest the fall in system frequency, in addition to governor primary control. Such interruptible loads and primary control are collectively called instantaneous frequency reserves.

Standard fixed speed induction generators contribute to the inertia of the power system because the stator is directly connected to the power system. Any change in power system frequency manifests as a change in the speed of stator-led rotating flux. Such speed changes are resisted by the rotating mass (generator rotor and the wind turbine rotor) leading to rotational energy transfer to the power system via the stator.

In modern variable-speed wind turbines, its rotational speed is normally decoupled from the grid frequency by the power electronic converter. Therefore variation in grid frequency don not alter the turbine output power. With high wind power penetration there is a risk that the power system inertial effect decreases, thus aggravating the grid frequency stability. The decrease of inertia effect on the grid may be even worse in power system with slow primary frequency response such as those large amount of hydropower, or in small power systems with inherent low inertia system such as islanded systems[18].

As the penetration of wind is expected to grow dramatically in the coming decade, researchers and vendors have sought improved designs to allow these technologies to better contribute to grid frequency regulation and stability. As noted above, most of the solutions proposed to date seek to mimic the inherent inertial response of traditional synchronous generators; i.e., they add a control loop that incrementally feeds or draws active power in response to a decline or rise in the time derivative of frequency. The control power required by this proposed additional loop comes predominantly either by varying the mechanical input power to a wind turbine, through change in its blade pitch or nominal rotational speed, or by drawing/feeding additional active power from/to the grid through the rotor side converter [19].

2.9. Wind probability density distribution

It is essential to assess wind energy potential of a site before any wind energy based system could be set up. Study of wind velocity regime over a period of time in a locality can really help to optimize the design of the wind energy conversion system by ensuring less energy generating costs. Wind velocity is generally recorded in a time-series format, which means wind velocity recorded over hourly basis in a day or over 24 h in a day.

To date, Weibull density function method is widely accepted for evaluating local wind load probabilities and is considered as a standard approach [20]. This method has a great flexibility and simplicity. However, the main limitation of the Weibull density function is its inability to accurately calculate the probabilities of observing zero or very low wind velocities [21]. Also Weibull two-parameter density function does not address the differences of wind velocity variation during the course of a day. Nevertheless, this statistical method is found to fit a wide collection of recorded wind data [22]. The Weibull wind velocity probability density function can be represented as [23]:

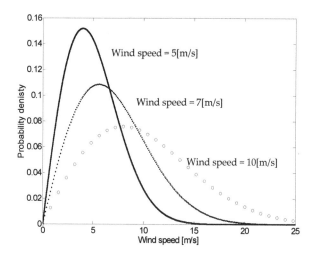

Figure 5. Probability distribution for the wind speed

$$f(\upsilon) = \frac{k}{c}\left(\frac{\upsilon}{c}\right)^{k-1} \exp\left[-\left(\frac{\upsilon}{c}\right)^{k}\right] \tag{3}$$

Where; $f(\upsilon)$ is the probability of observing wind velocity υ, c is the Weibull scale parameter and k is the dimensionless Weibull shape parameter. Basically, the scale parameter, c, indicates how 'windy' a wind site under consideration is, whereas the shape parameter, k, indicates how peaked the wind distribution is (that is, if the wind speeds tend to be very close to a certain value, the distribution will have a high k value and be very peaked). The probability distributions for low, medium and high wind speed locations are shown in Fig. 5. The figure demonstrate the difference in the probability distribution between high and low wind speeds.

3. Wind power impacts on the power system

Wind power has impacts on power system operational security, reliability and efficiency. Therefore, it is necessary to know the consequences of dynamic interaction between large scale wind farms and electrical power systems before incorporation of the wind farms into the grid. The electric power supply undergoes a change from a well-known and developed technology of conventional power plants to a partly unknown technology of wind power. High penetration of wind power could be managed through proper wind power plant interconnection, integration of the generation, transmission planning, and system operations.

Fig. 6 and 7 show impacts of wind power on power systems, divided in different time scales and width of area relevant for the studies. At the time of developing the standard IEC 61400-21: "Measurement and assessment of power quality characteristics of grid connected wind turbines", the wind turbines were mainly connected to the distribution grid, and the basic concern was their possible impact on the voltage quality and not on power system operation. This has changed with the development of large power rated wind farms that may form a significant part of the power system. In consequence, today's wind turbines are able to control the power (active and reactive) delivered both in transient and steady state, they can cope with power ramp requirements and they have low voltage ride through(LVRT) capability. They may even contribute to the primary frequency control, but then on the cost of dissipating energy [24]. These impacts can be categorized as follows:

- Short-term: impacts on the operational time scale (minutes to hours).

- Long-term: impacts on planning the transmission network and installed generation capacity for adequacy of power.

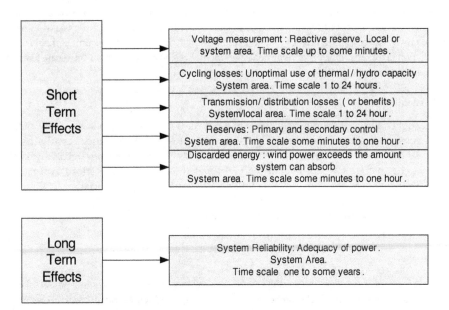

Figure 6. Power system impacts of wind power, causing integration costs [8]

For estimating the impacts, the different timescales involved usually mean different models that used in impact studies. This is why the impacts should be divided into three focus areas [23]:

3.1. Balancing

The first integration of renewable energy sources occurred on a small scale and at medium voltage. No extra measurements were taken for balancing the fluctuating power. However, as electricity grids are facing large scale integration of wind power, imbalances are occurring more frequently and are growing in magnitude. The first way to deal with the variable nature of wind power is to nominate the power day-ahead. Every balance responsible party (BRP) with wind turbines in his portfolio can nominate wind power output day-ahead by forecasting the predicted output. This system is extensively dealt with in the appendix concerning market mechanisms. Thus, the power generation is matched with the expected power demand. After this nomination, prediction errors lead to imbalances in the portfolio of each BRP. A particular BRP has the opportunity to balance its portfolio with different intra-day mechanisms (if available).

Again, the power generated is matched with the expected power demand. After gate closure, it can be expected that all remaining imbalances are dealt by the Transmission System Operator (TSO). Thus, prediction errors of one up to three hours ahead are to be balanced by the TSO. The TSO uses primary, secondary and tertiary reserves to balance the power imbalances as a result of prediction errors.

Wind power imbalances have two origins: prediction errors and inter-prediction deviations because the nominated power amount is for a fixed time period. Inter-prediction errors are related to the variable output of wind. Suppose the power is nominated on a basis of 15 min, wind power varies around the nominated value leading to inter- and intra-minute imbalances. Prediction errors result in positive or negative imbalances on a much longer timescale to even Inter-hour imbalances [24].

3.2. Adequacy of power

Total supply available during peak load situations (time scale: several years, and associated with static conditions of the system). The estimation of required generation capacity needs includes the system load demand and maintenance needs of production units (reliability data). The criteria that are used for the adequacy evaluation include the loss of load expectation (LOLE), the loss of load probability (LOLP) and the loss of energy expectation (LOEE), for instance. The issue is the proper assessment of wind power's aggregate capacity credit in the relevant peak load situations – taking into account the effect of geographical dispersion and interconnection [25].

3.3. Grid

The impacts of wind power on transmission depend on the location of wind power plants relative to load, and the correlation between wind power production and electricity consumption. Wind power affects power flow in the network. It may change the power flow direction, and reduce or increase power losses and bottleneck situations. There are a variety of means to maximize the use of existing transmission lines like use of online information (temperature, loads), FACTS, and wind power-plant output control. However, grid reinforcement may be necessary to maintain transmission adequacy and security. When determining adequacy of the grid, both steady-state load flow and dynamic system-stability analysis are needed. Different wind turbine types have different control characteristics, and consequently, also have different possibilities to support the system in normal and system-fault situations. For system stability reasons, operation and control properties will be required from wind power plants at some stage, depending on wind power penetration and power system robustness [25].

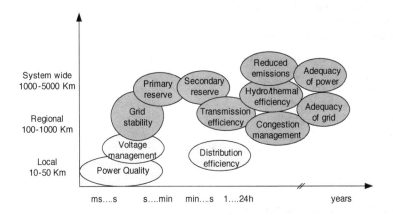

Figure 7. Impacts of wind power on power systems, divided in different time scales and width of area relevant for the studies[19]

4. Wind power integration solutions

Studies show that for an individual wind turbine, the variation in output is small for time-scales of less than a few seconds; for an individual wind farm, the variation in output is small for time scales of tens of seconds, due to the averaging of output of individual turbines across the wind farm; and for a number of wind farms spread out across a large area, such as a national grid system, the variation in output of all wind turbines is small for timescales from minutes to tens of minutes. The power produced from a large number of wind turbines will vary relatively less than the power produced from a single wind turbine due to the cancellation effect from the poor spatial correlation of the wind acting on each wind turbine.

To enhance the security of supply, new transmission and distribution grid codes specify technical requirements such as fault-ride through capability and frequency control of the electrical conversion systems of wind farms. Fault ride-through capability refers to the generators capabilities to remain connected to electricity networks at voltage levels below nominal. Active power control is closely related to frequency control and the wind farm shall have frequency control capabilities to ramp up and down the wind farm power station's active power output in accordance with the frequency/active power characteristic defined by the grid operator [26], [27].

When a power system is subjected to a sudden increase in reactive power demand following a system disturbance, the additional demand must be met by reactive power reserve carried by generators and compensators. If wind farms or other generation units are unable to withstand voltage drops for a limited time, they will disconnect from the system and then the reactive power supplied by these generators is lost that can entail load shedding or even a blackout, in the worst case [28]. To ensure the voltage recovery the wind-turbine generators must remain connected to the system to provide reactive power support after the fault clearance. For many wind turbine manufacturers these are very costly and challenging requirements. In some cases extensive modifications to the electrical system of the turbines are necessary [29].

Achieving reliable operation at greatly reduced voltage levels is proving problematic. A particular problem regarding power converter-based wind-turbine generators is that conventional controllers for power converters designed for reliable operation around nominal voltage levels will not work as designed during low network voltages that can occur during a fault. A consequence of this is greatly increased converter currents, which may lead to converter failure. New controller design strategies have been proposed for power converter-based wind turbine generators aiming to maintain converter currents within their design limits, even at greatly reduced voltage levels, in order to enhance the wind-turbine generators' fault ride-through capability [30]. With the increasing penetration of power converter-based wind turbine generators the rotational speed of the wind turbines are decoupled from the grid leading to a reduction of inertia in the grid. The lower the inertia of a system, the more and faster the frequency will change with variations in generation or load. In order that a variable-speed wind turbine to contribute to the system inertia and the frequency control as a result it has been proposed in [31] an additional control loop in the power electronic converter which connects the turbine inertia directly to the grid so that the wind turbine will be able to increase its power supplied to the grid during a drop in the grid frequency.

Whilst the wind farms are considered like other generating facilities by some grid operators and as such they are requested to participate in the system frequency and voltage compensation, the wind power sector claims for less strict requirements which imposes unnecessary burden and cost on manufacturers. The wind power sector calls on an overall economically efficient solution where the primary and secondary control should be provided by conventional power plants with the wind farms providing such service only in cases where limits in existing reserves are foreseen, and reactive power compensation provided by FACTS devices directly installed in the transmission network [31].

5. Conclusion

The complexity of power systems has increased in recent years due to the operation of existing transmission lines closer to their limits due to the increased penetration of new types of generators that have more intermittent characteristics and lower inertial response, such as wind generators. This changing nature of a power system has considerable effect on its dynamic behaviors resulting in power swings, dynamic interactions between different power system devices and less synchronized coupling.

Understanding and quantifying the impacts of wind farms on utility systems is a critical first step in identifying and solving problems. The design and operation of the wind plant, the design and operation of the power system, and the market rules under which the system is operating influence the situation. A number of steps can be taken to improve the ability to integrate increasing amounts of wind capacity on power systems such as improvements in wind-turbine and wind- farm models, improvements in wind-farm operating characteristics, improvements in the flexibility of operation of the balance of the system, carefully evaluating wind-integration operating impacts, incorporating wind-plant output forecasting into utility control-room operations, making better use of physically available transmission capacity, upgrading and expanding transmission systems, developing well-functioning hour-ahead and day-ahead markets and expanding access to those markets, adopting market rules and tariff provisions that are more appropriate to weather-driven resources, and consolidating balancing areas into larger entities or accessing a larger resource base through the use of dynamic scheduling or some form of area control error (ACE) sharing.

As additional integration studies and analyses are conducted around the world, it is expected that more researches will be valuable as wind penetration increases. And with the large increase in installing wind farms, actual practical experience will also contribute strongly in our understanding of the effects that arise from the increasing installation of wind farms on the system as well as on ways that the impacts of wind's variability and uncertainty can be treated in an inexpensive manner.

Author details

Ahmed G. Abo-Khalil

Electrical Engineering Department, Majmaah University, Saudi Arabia

References

[1] M.J. Hossain, H.R. Pota, M.A. Mahmud, and R. A. Ramos, "Impacts of large-scale wind generators penetration on the voltage stability of power systems, " IEEE Power and Energy Society General Meeting, PP: 1-8, July 2011.

[2] World Wind Energy Association WWEA, "World Wind Energy Report 2010".

[3] Renewable Energy Policy Network for the 21Century, "Renewables 2011: Global Status Report," 2011.

[4] Eolic Energy News, "Spain becomes the first European wind energy producer after overcoming Germany for the first time", Dec. 2010.

[5] Renewable Energy Research Laboratory, "Wind Power: Capacity Factor, Intermittency," University of Massachusetts, 2008.

[6] American Wind Energy Association "20% Wind Energy by 2030" 2009.

[7] H. Holttinen *et al*, " State-of-the-art of design and operation of power systems with large amounts of wind power summary of IEA wind collaboration *EWEC2007*," *Proc. European Wind Energy Conf. May 2007.*

[8] Deutsche Energie-Agentur Dena (DENA), "Planning of the grid integration of wind energy in Germany onshore and offshore up to the year 2020," 2005.

[9] G. Strbac, A. Shakoor, M. Black, D. Pudjianto, and T. Bopp, "Impact of wind generation on operation and development of the future UK electricity systems ,"*Electr. Power Syst. Res., 2007.*

[10] T. Ackermann, J. R. Abbad, I. M. Dudurych, I. Erlich, H. Holttinen, J. R. Kristoffersen, and P. E. Sorensen, "European balancing act," *IEEE Power Energy Mag.*, vol. 5, no. 6, pp. 90–103, 2007.

[11] Z. Lubosny, Wind Turbine Operation in Electric Power Systems, Springer-Verlag Berlin, ISBN 3-540-40340-X.

[12] Z. Chen, " Issues of Connecting Wind Farms into Power Systems," IEEE/PES Transmission and Distribution Conference & Exhibition: Asia and Pacific Dalian, China, 2005.

[13] Ö.S. Mutlu, E. Akpınar, and A. Balikci," Power Quality Analysis of Wind Farm Connected to Alacati Substation in Turkey," Renewable Energy Journal, vol. 34 pp. 1312–1318, 2009.

[14] D. Weissera, and R. S. Garcia " Instantaneous wind energy penetration in isolated electricity grids: concepts and review," Elsevier, Renewable Energy Journal, pp. 1299-1308, 2005.

[15] L. Munteanu, A. L. Bratcu, N.-A. Cutulilis, and E. Ceanga, Optimal Control of Wind Energy Systems, Springer, ISBN 978-1-84800-080-3.

[16] I. de Alegria, J. Andreu, J. Martin, P. Ibanez, J. Villate, and H. Camblong, "Connection requirements for wind farms: A survey on technical requierements and regulation," Renewable and Sustainable Energy Reviews, vol. 11, no. 8, pp. 1858–1872, 2007.

[17] J. Hethey and M. S. Leweson, "Probabilistic Analysis of Reactive Power Control Strategies for Wind Farms," M.Sc. thesis, Technical University of Denmark, 2008.

[18] G. C. Tarnowski, P. C. Kjaer, P. E. Sorensen, and J. Ostergaard, " Study on variable speed wind turbines capability for frequency response", in *Proc. European Wind energy Conference EWEC*, 16-19 March, Marseille, France, 2009.

[19] A. Keane, M. Milligan, C. D'Annunzio, C. Dent, K. Dragoon, B. Hasche, H. Holttinen, N. Samaan, L. Söder, M. O'Malley, "Calculating the capacity value of wind," IEEE Transactions on Power Systems, vol. 26, no. 2, pp. 564-572, 2011.

[20] L. Lu, H. Yang H, and J. Burnett, "Investigation on wind power potential on Hong Kong—an analysis of wind power and wind turbine characteristics," Renewable Energy Journal, vol. 27, pp.1–12, 2002.

[21] D. Weisser, " A wind energy analysis of Grenada: an estimation using the Weibull density function," Renewable Energy Journal, vol. 28, pp. 1803-1812, 2001.

[22] I.Y.F Lun and J. C. Lam (2000)," A study of Weibull parameters using long-term wind observations," Elsevier, Renewable Energy Journal vol. 20, No. 2, 145–53, 2000.

[23] C. Justus, W. Hargraves, A. Mikhail, and D. Graber (1978), " Methods for estimating wind speed frequency distributions," J. Appl. Meteorol., vol. 17, 350– 353.

[24] H. Holttinen et al, "Design and operation of power systems with large amounts of wind power" Final report, IEA WIND Task 25, 2006-2008.

[25] H. Holttinen, P. Meibom, A. Orths, B. Lange, M. O'Malley, J. O. Tande, A. Estan-queiro, E. Gomez, L. Söder, G. Strbac, J. C. Smith, F. van Hulle, "Impacts of large amounts of wind power on design and operation of power systems, results of IEA collaboration," Journal of Wind Energy, 2008.

[26] R. Leão, T. Degner and F. Antunes, "An overview on the integration of large-scale wind powerinto the electric power system," Dept. of electrical engineering, Federal University of Ceará, Brazil, 2007.

[27] M. Dahlgren, H. Frank, M. Leijon, F. Owman, L. Walfridsson, "WindformerTM Wind power goes large-scale", ABB review, no. 3, pp. 31–37, 2000.

[28] Prabha Kundur, Power system stability and control, ISBN 0-07-035958-X.

[29] S. Repo, "On-line Voltage Stability Assessment of Power System- An Approach of Black-box Modelling," PhD Thesis, Aalborg University, 2001.

[30] A. Mullane et al., "Wind-turbine fault ride-through enhancement," IEEE Trans. Power Systems, vol. 20, no. 4, pp. 1929-1937, 2005.

[31] J. C. Smith, "Winds of change: Issues in utility wind integration," vol. 3, pp. 20–25, Nov. 2005.

Chapter 9

Study for Wind Generation and CO_2 Emission Reduction Applied to Street Lighting – Zacatecas, México

Francisco Bañuelos-Ruedas,
César Ángeles-Camacho,
Guillermo Romo-Guzmán and
Manuel Reta-Hernández

Additional information is available at the end of the chapter

1. Introduction

The widespread concern in developed and developing countries to generate clean and sustainable energy, has led to search for alternative sources for non polluting power generation such as wind power. Although electric power generating costs by harnessing the wind resource are still higher than production with conventional plants, the difference is being reduced, depending on the system capacity. Integrating wind power systems to distributed generation scheme, the efficiency of transmission and distribution may increase.

An specific application of wind generation is to provide electric energy for street lighting in cities or towns close to wind farms. In Mexico, the cost of wind power electricity may satisfy the demand in public lighting with acceptable cost per kWh.

Since ancient times, the wind has been used for various purposes, including navigation, grain mills and irrigation. It was until the early twentieth century when wind power started his application in electric generation. It was more expensive, though, to produce electricity from wind power than with conventional fossil fuels plants. In recent decades, the technology development to harness the wind resource has accelerated, and today many countries use the wind resource on a large scale at competitive costs. It started with small generators using a few watts of power, and currently there are up to 5 MW wind turbine generators with possible capacity of 7, 10 and 20 MW for the coming years [1]. Figure 1 shows the growth of wind turbines related to their installation heights.

This chapter provides a study concerning an estimation of wind resources and the possibility of supplying electricity for street lighting from wind farms in the state of Zacatecas, Mexico. It also presents a summary of environmental impact concerning the tons of CO$_2$ not released to the environment using this type of generation.

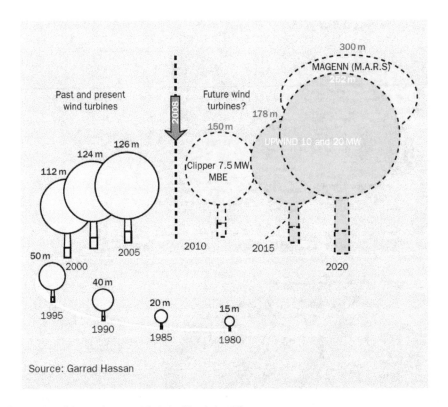

Figure 1. Growth in size of commercial wind turbine designs [1]

2. Wind resource in Mexico

According to a wind resource evaluation performed in 1995 by Schwartz and Elliott [2], México has interesting regions with wind capacity to produce electric energy. Figure 2 a) shows estimated utility-scale areas in Oaxaca (Istmo de Tehuantepec), Veracruz, Tamaulipas, Yucatan, Quintana Roo, Baja California and Zacatecas, with wind power classes from 3 to 5. Figure 2 b) shows wind capacity for rural-scale areas for the rest of the country with wind power classes from 1 to 4.

Mexico—Preliminary Wind Resource Map for Utility-Scale Applications

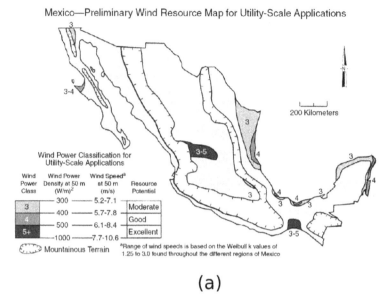

(a)

Mexico—Preliminary Wind Resource Map for Utility-Scale Applications

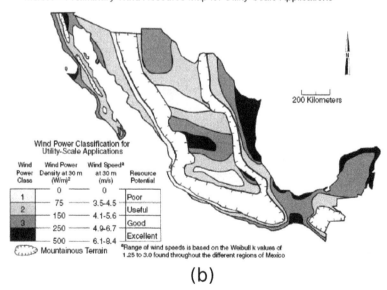

(b)

Figure 2. Preliminary wind resource of Mexico estimated by Schwartz and Elliott [2], a) for utility-scale applications; b) for rural-scale applications.

In 2007, Klapp, Cervantes-Cota and Chavez [3] published estimated wind power data for Mexico, showing potential wind power capacity in MW (Figure 3). Some of the more studied areas shown are Zacatecas (400 MW), Oaxaca and Chiapas (2000 MW), and Baja California (100 MW). Some of the less studied areas are Tamaulipas (700 MW), Veracruz, Hidalgo and Puebla (600 MW), Baja California Sur (50 MW), Quintana Roo and Yucatán (800 MW), Chihuahua (50 MW), and Sinaloa (100 MW). The estimated capacity factors go from 18 % to 30%, and 50% in Oaxaca (Istmo de Tehuantepec).

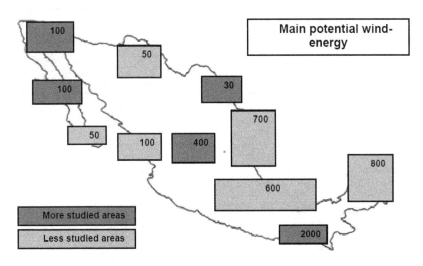

Figure 3. Estimated wind potential areas in Mexico, in MW, shown by Klapp, Cervantes-Cota and Chavez [3]

The wind power prospective in Mexico 2011-2025 reported by SENER [4]-[5], estimates a wind resource of 71,000 MW considering capacity factors between 20% and 30% (Table 1). For capacity factors between 30% and 35% the estimated wind potential is around 11,000 MW. For capacity factors beyond 35%, the estimated wind potential is 5,235 MW.

Capacity factor (%)	Land (%)	Estimated Wind Capacity (MW)
20-25	56.7	40,268.0
25-30	27.5	19,535.0
30-35	8.4	5,961.0
35-40	3.5	2,500.0
> 40	3.9	2,735.0
Total	100	71,000

Table 1. Estimated wind energy potential in Mexico, by SENER [4,5]

3. Wind power plants in Mexico

The significant wind power plants installed in Mexico have been developed during the last seven years. Almost all plants have been installed in Oaxaca, due to its high wind potential, although other regions are being considered. Some other areas are currently being monitored for possible wind exploitation. Table 2 shows the wind power projects developed until 2010, according to SENER [5] and AMDEE [6].

At the end of 2011, the wind capacity installed in Mexico reached 873 MW (position 19 in the global ranking, according to GWEC [7]). There are some other projects in planning stage to be constructed during the next three years, expecting a total capacity of 6,792.7 MW at the end of 2014 [6].

Project	Location	Developer	Date of commercial operation	Capacity (MW)
La Venta	Oaxaca	CFE	1994	1.6
La Venta II	Oaxaca	CFE	2006	83.3
Parques Ecológicos de México	Oaxaca	Iberdrola	2009	79.9
Eurus, phase 1	Oaxaca	Cemex/Acciona	2009	37.5
Eurus, phase 2	Oaxaca	Cemex/Acciona	2010	212.5
Gobierno Baja California	Baja California	GBC/Turbo Power Services	2010	10
Bii Nee Stipa I	Oaxaca	Cisa-Gamesa	2010	26.35
La Mata - La Ventosa	Oaxaca	Eléctrica del Valle de México (EDF-EN)	2010	67.5
		Total		518.63

Table 2. Wind power plants in operation in Mexico at the end of 2010 [5,6]

The Energy Department, SENER [4], expects a continuos and sustained growing in all renewable energy sectors for electric energy production, predicting a total wind capacity of 11,703 MW at the end of 2024. Table 3 shows the distribution of expected electricity production with all types of renewable resources. It can be observed that wind power capacity will be the second resource (39.6%), only after hydroelectric power. The total predicted value is 31,854 MW.

Resource	MW	%
Hydroelectric < 30 MW	1,348	4.2
Hydroelectric > 30 MW	14,657	46.2
Solar PV (Photovoltaic)	1,942	6.1
Solar CSP (Concentrated Solar Power)	69	0.2
Bioenergy	905	2.9
Geothermal power	1,230	3.6
Wind power	11,703	39.6
Total	31,854	100.0

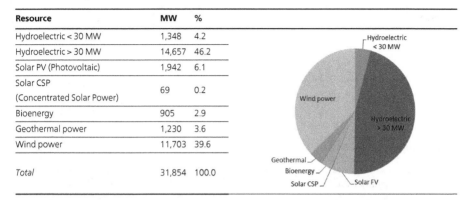

Table 3. Predicted values of electricity production from renewable energy resources in Mexico in 2024 [4]

4. Wind resource estimation

The construction feasibility of any wind project requires the fulfillment of several points, such as:

a. Selection of site.

b. Wind speed and wind direction monitoring.

c. Wind rose description.

d. Electric network close to the site.

e. Environmental studies.

f. Economical / social studies.

g. Geographical access.

h. Legal permits.

i. Wind turbine/generator electro–mechanical modeling.

j. Technical /economical analysis of the wind plant.

The impact of wind generation projects must be evaluated by two fundamental factors: enviromental issues and power electric grid characteristics.

In the first case, it can be outlined the following impacts [8]:

• Atmosphere.

• Effects on flora.

• Effects on birds.

- Visual.

- Noise.

The impact of wind power generation on the power electric grid may be measured on short, middle and large periods of time, taking into account factors as:

- Level of penetration.

- Capacity of electric grid.

- Structure of power generation.

The level of possible penetration can be determined by (a), (b) y (c) described above in the wind resource estimation.

The wind resource estimation of the site requires monitoring of several climate variables as wind speed, wind direction, temperature and atmospheric pressure taken, at certain hight, every two minutes during, at least, twelve consecutive months. All data obtained is statistically processed through specialized computational tools to obtained plots and characteristics curves [9]-[10] like wind rose wind speed–frequency distribution curve. The wind resource of a region (wind map) is obtained later, considering the data of several sites.

Figures 4 and 5 show the wind rose and wind speed-frequency distribution curve obtained in a monitoring station [11]-[12].

Figure 6 shows the wind maps of a region, indicating a) annual average values of wind speed and, b) annual average values of wind power density.

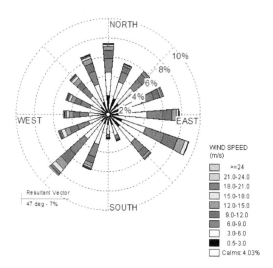

Figure 4. Wind rose obtained in a selected site. Zacatecas, Mexico

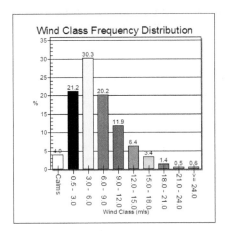

Figure 5. Wind speed-Frequency distribución curve obtained in a selected site. Zacatecas, Mexico

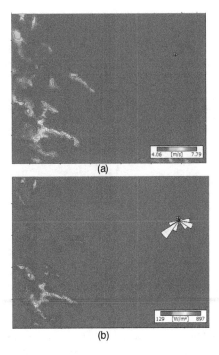

Figure 6. Wind map of a selected region in Zacatecas, Mexico, obtained with WAsP© software, showing (a) annual average wind speed, in m/s, and (b) annual average power density, in W/m².

With the wind maps and their corresponding wind roses, the next step to calculate the wind plant is to select the specific area within the studied region that fulfill the environmental requirements.

A wind power plant or wind farm, usually have several wind turbines distributed in the selected area in such a way the available wind resource can be well exploited. The proper turbines arrangement is obtained, generally, by using computational tools and digital simulators, all in compliance with existing national and international regulations.

The electricity produced by the wind turbines in large power plants is fed to power systems through electric transformers and power electronic controllers [13]-[14]. Once the electric energy is sent to the network, it can be applied to different electric loads. Figure 7 shows a general scheme diagram of electricity production and consumption using wind power.

The present document proposes to apply the electric energy produced by the wind turbines in public street lighting or in municipal water pumping. Both loads have high tariff rates in Zacatecas.

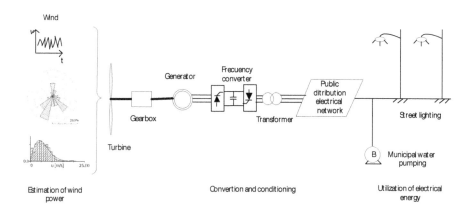

Figure 7. General scheme of electricity production and consumption using wind power.

5. Electric energy demand in Mexico and Zacatecas state

Mexico had during 2011 a total consumption of 186,638,847 MWh of electric energy, with 4.13% consumed in the public sector (7,706,706 MWh). From the total energy demanded by the public sector, 61.55% was consumed in street lighting and 38.44% of was consumed in water pumping. Table 4 shows the electric energy values demanded by the whole country and by Zacatecas state.

Sector	National		Zacatecas state	
	MWh	%	MWh	%
Residential	48,700,399	26.09%	480,403	18.72%
Commercial	12,991,134	6.96%	121,654	4.74%
Agriculture	8,599,593	4.61%	503,559	19.63%
Mid-size industry	70,024,362	37.52%	304,389	11.86%
Large-size industry	38,616,680	20.69%	937,553	36.54%
Public services	7,706,706	4.13%	218,234	8.51%
Total	186,638,874	100.00%	2,565,792	100.00%
Public services				
Municipal water pumping	2,962,516	1.59 %	113,460	4.42%
Street Lighting	4,744,190	2.54%	104,774	4.08%
Total	7,706,706	4.13%	218,234	8.51%

Table 4. Electric energy consumption during 2011 in Mexico and in Zacatecas state [15]

During 2008 the total electric energy consumption in Zacatecas state was 1,726,935 MWh. During 2011, the total consumption increased to 2,565,792 MWh. There have been increases in consumption tariffs for public lighting services and municipal water pumping in the state. Table 5 shows the sales reported by electric utility in Zacatecas state in public sector.

Sector	No of customers	Energy sold (MWh)	Energy sold (%)	Average price ($ pesos / kWh)	Thousands of pesos
Municipal water pumping (Tariff 6)	1,369	113,460	4.42	1.40	162,901
Street Lighting (Tariff 5A)	10,078	104,774	4.08	2.10	220,253

Table 5. Sales reported by electric utility in Zacatecas state during 2011 [15].

6. Case study-wind electric energy production for public street lighting

6.1. Capacity of wind power plant determination

According to electricity utility, CFE, energy sales for public street lighting in Zacatecas state during 2011 were 104,774 MWh, representing a continuous operating capacity of 11.96 MW. The total energy sales were 2,565,792 MWh, representing, in average, a continuous generation capacity of 292.89 MW for all services.

The electricity consumption for public lighting in the most important city of Zacatecas state, is in average 1,100 MWh per month, representing approximately a wind power plant generation capacity of 1.52 MW with a capacity factor of 30%. This means a total wind power plant capacity of 5.09 MW. The wind power plant may be build with three sets of wind turbines/generators of 2 MW located in a region near to the mentioned city with the required wind capacity [16,17].

6.2. Economic analysis of wind generation costs

Table 6 shows comparison of electricity generation costs using different technologies [18]. It can be seen that wind power generation is a competitive choice in 2011, compared to diesel and steam (oil) technologies. Certainly, it is not a good choice comparing to the other technologies costs outlined in the table, but if other issues as environmental and healthy problems are taken into account, wind power technology is not a bad selection.

Technology	Year			
	2008	2009	2010	2011
Diesel	7.85	8.27	15.91	16.58
Steam (oil fuel)	1.58	1.50	1.79	2.01
Wind power	0.74	0.69	1.02	1.84
Nuclear	1.12	1.05	1.97	1.26
Dual (Coal and oil)	1.10	0.98	0.90	0.96
Turbo Gas and Combined Cycle	1.38	0.87	0.90	0.94
Geothermal	0.59	0.48	0.47	0.56
Hydroelectric Generation	0.49	0.63	0.44	0.51

Generation cost includes:
• Salaries and employee benefits
• Energy and power purchased
• Maintenance and general services contract
• Maintenance and materials consumption
• Taxes and duties
• Cost of labor obligations
• Depreciation
• Indirects costs
• Development and financial cost.
• Other expenses

Table 6. Electricity generation costs in $ pesos / kWh, using different technology [18].

6.3. Environmental impact

To estimate the environmental impact let us considered the average per month in the operation of the proposed 6 MW wind power plant capacity mentioned in the previous section, the total power delivered during this period will be 1.2 GWh. If contaminants emission of a coal power plant has a rate of 1,058.2 tons of CO_2/GWh, and a rate of 7.4 tons of CO_2/GWh for wind power plant, thus the operation of the wind power plant represents a reduction of 1260.96 tons of CO_2 per month, and 15,131.52 tons of CO_2 per year. Table 7 presents a comparison of CO_2 emissions in conventional power plants and in a wind farm that supplies 1.2 GWh per month. The emission factors are based on references [19-21].

Source	Capacity in GWh per month	CO_2/GWh (Tons)	Total CO_2 emitted (Tons)
Coal	1.2	1,058.20	1,269.84
Oil	1.2	820	984.00
Natural gas	1.2	524	628.80
Wind energy	1.2	7.4	8.88

Table 7. Comparison of CO_2 emission per month from fossil fuel plants and wind power plant of 1.2 GWh

All electric energy demanded per year for public street lighting in Zacatecas state (104,774 MWh), representing a continuous operation capacity of 11.96 MW, could be supplied by a wind power plant with a total capacity of 40 MW, capacity factor 30%. Table 8 shows the CO_2 emissions produced per year in conventional power plants and in a wind farm that supplies 104.77 GWh per year.

Source	Capacity in GWh per month	CO_2/GWh (Tons)	Total CO_2 emitted (Tons)
Coal	104.774	1,058.20	110,871.85
Oil	104.774	820	85,914.68
Natural gas	104.774	524	54,901.58
Wind energy	104.774	7.4	775.33

Table 8. Comparison of CO2 emission per year from fossil fuel plants and wind power plant of 104.77 GWh

In Tables 7 and 8 can be observed the big differences in CO_2 emissions by using fossil fuel and wind energy technologies applied to public street lighting in Zacatecas. This is a simple example of the benefits that can be obtained by using wind energy.

7. Conclusions

Monitoring and estimation of wind resource of an specific site are fundamental tasks to start a wind power plant project. They determine if the site has the minimum requirements to exploit the wind resource. Once the wind capacity is evaluated, the next step is to establish the electric energy demand.

In this chapter it is discussed how the electric energy demand for public street lightning and water pumping in the State of Zacatecas, Mexico, could be supplied by wind energy. Nowadays, the electricity produced by wind power plants has reached competitive costs that may be applied in public street lighting, besides the reduction of CO_2 emitted to the atmosphere and its corresponding carbon bonuses. It is concluded that is widely recommended to apply the wind energy production in public street lightning for Zacatecas State, even tough is necessary to complete more extensive environmental and grid impact studies.

Acknowledgments

The authors wish to give their acknowledgments to Universidad Autónoma de Zacatecas (UAZ) and Instituto de Ingeniería, UNAM. F. Bañuelos-Ruedas, G. Romo-Guzmán y M. Reta-Hernandez thank the support to Unidad Académica de Ingeniería Eléctrica, UAZ, and C. Angeles-Camacho thanks the support given by Program for Research and Technology Innovation Project (PAPIIT), through project IN151510, to complete the present chapter.

Author details

Francisco Bañuelos-Ruedas[1], César Ángeles-Camacho[2], Guillermo Romo-Guzmán[1] and Manuel Reta-Hernández[1]

1 Universidad Autónoma de Zacatecas, México

2 Instituto de Ingeniería de la UNAM, México

References

[1] European Wind Energy Association. EWEA: Wind energy – the facts, Executive summary.http://www.ewea.org/fileadmin/ewea_documents/documents/publications / WETF/1565_ExSum_ENG.pdf. (accessed 30 July 2012).

[2] Schwartz MN and Elliott DL. Mexico wind resource assessment project. DOE/NREL Report No. DE95009202, National Renewable Energy Laboratory, Golden, Colorado, March 1995.

[3] Klapp J, Cervantes-Cota J and Chávez L J., editors. Towards a cleaner planet. Energy for the future. New York: Springer; 2007.

[4] Secretaría de Energía. SENER. Prospectiva de energías renovables 2011 – 2025. México: SEDE; 2011.

[5] Secretaría de Energía. SENER. Prospectiva del sector eléctrico 2011 – 2025. México: SEDE; 2012.

[6] Asociación Mexicana de Energía Eólica. AMDEE: Proyectos eólicos en México. http://amdee.org/Recursos/Proyectos_en_Mexico. (accessed 3 July 2012).

[7] Global Wind Energy Council. GWEC: GlobalWind Statistics 2011. http://www.gwec.net/ (accessed 4 July 2012).

[8] Mur, J. Curso de energía eólica. España. Departamento de Ingeniería Eléctrica de la Universidad de Zaragoza.http://www.joaquinmur.eu/manualEolico.pdf (accessed 16 June 2012)

[9] WRPLOT View program Version 5.9 by Lakes Environmental. Canada. © 1998-2008.

[10] Wind Atlas Analysis and Application Program (WAsP©) by Risø National Laboratory, Technical University of Denmark (DTU), Roskilde, Denmark. © 1987-2007.

[11] Medina, G., Reporte Agrometereológico August 2005 - July 2006, INIFAP Centro de Investigación Regional Norte Centro, Campo Experimental Zacatecas. Boletines informativos No. 15-26. México.

[12] Instituto de Investigaciones Eléctricas. IIE. Información anemométrica de la Estación Cieneguillas, Zacatecas. 2005-2006. http://planeolico.iie.org.mx. (accessed 24 March 2008.

[13] Ackerman T., editor. Wind Power in Power Systems. 2nd Ed. USA: John Wiley and Sons; 2005.

[14] Wildi T. Electrical Machines. Drives and Power Systems 6th Ed. USA: Prentice Hall; 2006.

[15] Comisión Federal de Electricidad CFE. Estadísticas. http://app.cfe.gob.mx/ Aplicaciones/QCFE/EstVtas/Default.aspx. (accessed 12 July 2012)

[16] Reta-Hernandez M, Soto CE, De la Torre J, Ibarra S, Álvarez JA, Romo-Guzmán G, Bañuelos- Ruedas F, Ochoa Ortiz CA, Martínez, AE, Medina G. and Rumayor A F. Resultados preliminares de la evaluación del recurso eólico en varios sitios del Estado de Zacatecas. Presented in XXXI Semana nacional de energía solar. Zacatecas, México. October 2007.

[17] Ángeles-Camacho C, Bañuelos-Ruedas F. and Badillo-Fuentes JF. El recurso eólico en el Estado de Zacatecas: Características del viento en 36 localidades. México: II-UNAM; 2011.

[18] Comisión Federal de Electricidad. CFE: Costos de generación por tecnología. http://www.cfe.gob.mx/QuienesSomos/queEsCFE/Documents/2012/Administracion/Costo-degeneracionportecnologia2002_2011.pdf. (accessed 12 July 2012).

[19] Sovacool BK. Valuing the greenhouse gas emissions from nuclear power: A critical survey. Energy Police. 2008; 36(8) 2940-2953.

[20] Merino L. Energías renovables. (Colec. Energías renovables para todos). España. Haya   Comunicación.http://www.energiasrenovables.com/Productos   /pdf/cuaderno_GENERAL.pdf (accesed 18 June 2012).

[21] Spadaro V, Langlois L and Hamilton B. Greenhouse gas emissions of electricity generation chains: assessing the difference. IAEA. Bull 2000; 42(2) 19-24.

Permissions

The contributors of this book come from diverse backgrounds, making this book a truly international effort. This book will bring forth new frontiers with its revolutionizing research information and detailed analysis of the nascent developments around the world.

We would like to thank S. M. Muyeen, Ahmed Al-Durra and Hany M. Hasanien, for lending their expertise to make the book truly unique. They have played a crucial role in the development of this book. Without their invaluable contribution this book wouldn't have been possible. They have made vital efforts to compile up to date information on the varied aspects of this subject to make this book a valuable addition to the collection of many professionals and students.

This book was conceptualized with the vision of imparting up-to-date information and advanced data in this field. To ensure the same, a matchless editorial board was set up. Every individual on the board went through rigorous rounds of assessment to prove their worth. After which they invested a large part of their time researching and compiling the most relevant data for our readers. Conferences and sessions were held from time to time between the editorial board and the contributing authors to present the data in the most comprehensible form. The editorial team has worked tirelessly to provide valuable and valid information to help people across the globe.

Every chapter published in this book has been scrutinized by our experts. Their significance has been extensively debated. The topics covered herein carry significant findings which will fuel the growth of the discipline. They may even be implemented as practical applications or may be referred to as a beginning point for another development. Chapters in this book were first published by InTech; hereby published with permission under the Creative Commons Attribution License or equivalent.

The editorial board has been involved in producing this book since its inception. They have spent rigorous hours researching and exploring the diverse topics which have resulted in the successful publishing of this book. They have passed on their knowledge of decades through this book. To expedite this challenging task, the publisher supported the team at every step. A small team of assistant editors was also appointed to further simplify the editing procedure and attain best results for the readers.

Our editorial team has been hand-picked from every corner of the world. Their multi-ethnicity adds dynamic inputs to the discussions which result in innovative outcomes. These outcomes are then further discussed with the researchers and contributors who give their valuable feedback and opinion regarding the same. The feedback is then collaborated with the researches and they are edited in a comprehensive manner to aid the understanding of the subject.

Apart from the editorial board, the designing team has also invested a significant amount of their time in understanding the subject and creating the most relevant covers. They scrutinized every image to scout for the most suitable representation of the subject and create an appropriate cover for the book.

The publishing team has been involved in this book since its early stages. They were actively engaged in every process, be it collecting the data, connecting with the contributors or procuring relevant information. The team has been an ardent support to the editorial, designing and production team. Their endless efforts to recruit the best for this project, has resulted in the accomplishment of this book. They are a veteran in the field of academics and their pool of knowledge is as vast as their experience in printing. Their expertise and guidance has proved useful at every step. Their uncompromising quality standards have made this book an exceptional effort. Their encouragement from time to time has been an inspiration for everyone.

The publisher and the editorial board hope that this book will prove to be a valuable piece of knowledge for researchers, students, practitioners and scholars across the globe.

List of Contributors

Adnan Sattar, Ahmed Al-Durra and S.M. Muyeen
Electrical Engineering Department, The Petroleum Institute, Abu Dhabi, UAE

Tárcio A. dos S. Barros and Ernesto Ruppert Filho
Universidade Estadual de Campinas-UNICAMP, Faculdade de Engenharia Elétrica e Computação-FEEC, DSCE, Campinas, Brazil

Alfeu J. Sguarezi Filho
Universidade Federal do ABC-UFABC, Santo André, Brazil

C. E. Capovilla, A. J. Sguarezi Filho and I. R. S. Casella
Universidade Federal do ABC - UFABC, Brazil

E. Ruppert
Universidade Estadual de Campinas - Unicamp, Brazil

Ivan Jorge Gabe
Federal Institute of Rio Grande do Sul-IFRS, Farroupilha, Brazil

Humberto Pinheiro and Hilton Abílio Gründling
Federal University of Santa Maria, Santa Maria, Brazil

Roberto Daniel Fernández
National University of Patagonia San Juan Bosco, Argentina

Pedro Eugenio Battaiotto
National University of La Plata, Argentina

Ricardo Julián Mantz
National University of La Plata and Scientific Investigation Comission of Buenos Aires State (CICpBA), Argentina

E. Barrios-Martinez and C. Angeles-Camacho
Instituto de Ingeniería, Universidad Nacional Autónoma de México, UNAM, México

L.M. Castro
Universidad Michoacana de San Nicolás de Hidalgo, UMSNH, Michoacán, México

C.R. Fuerte-Esquivel
Instituto de Ingeniería, Universidad Nacional Autónoma de México, UNAM, México
Universidad Michoacana de San Nicolás de Hidalgo, UMSNH, Michoacán, México

Tamer A. Kawady and Ahmed M. Nahhas
Electrical Engineering Department, Umm Al-Qura University, Makkah, Saudi Arabia

Ahmed G. Abo-Khalil
Electrical Engineering Department, Majmaah University, Saudi Arabia

Francisco Bañuelos-Ruedas, Guillermo Romo-Guzmán and Manuel Reta-Hernández
Universidad Autónoma de Zacatecas, México

César Ángeles-Camacho
Instituto de Ingeniería de la UNAM, México

Printed in the USA
CPSIA information can be obtained
at www.ICGtesting.com
JSHW011404221024
72173JS00003B/413

9 781632 391926